W9-ALL-807

636.3 $6.95
Be

Belanger, Jerry
    Raising milk
goats the modern way

DATE DUE

MY 15 91	MAR 27 15	AP 09 03
DE 19 92	APR 28	SE 1 03
OC 3 '92	JUN 07	OC 08 03
NO 10 '92	AUG 21	OC 20 04
DE 29 '92	JUL 17	AP 20 05
MY 29 '93	OCT 21	OC 17 05
SE 7 '93	AUG 11	OC 25 06
SE 30 '93	NO 1 '99	MY 2
SEP 05 94	DE 07 '02	AU 18 22
JAN 27 95	JA 13 03	
MAR 06 95	MR 03 03	
	MR 17 03	

EAU CLAIRE DISTRICT LIBRARY

DEMCO

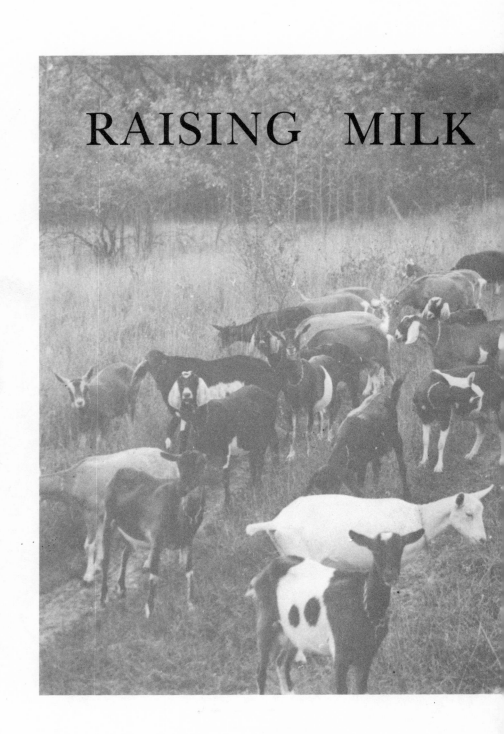

RAISING MILK

GOATS

THE MODERN WAY

Jerry Belanger

Editor, *Countryside and Small Stock Journal*

A Garden Way Publishing Book

Storey Communications, Inc.
Pownal, Vermont 05261

72844

EAU CLAIRE DISTRICT LIBRARY

Acknowledgments

This book couldn't have been written without all those beginning goat raisers who have written to *Countryside & Small Stock Journal* asking basic questions—as well as some pretty technical ones—nor could it have been written without the more experienced goat raisers who helped answer those questions.

It couldn't have been written without Judith Osborn Kapture, the dairy goat editor of *Countryside*. With her broad span of experience, meticulous and exhaustive research, professional reporting and genuine concern for the dairy goat and its owner, she, in my opinion, has done more to further the dairy goat in this country than any other individual.

And finally, this book would be nothing more than a rewrite of other people's experiences if it weren't for Calpurnia—my first goat—and all the others who followed.

J. B.

Copyright © 1975 by Storey Communications, Inc.

The name Garden Way Publishing is licensed to Storey Communications, Inc. by Garden Way, Inc.

All rights reserved. No part of this book may be reproduced without written permission from the publisher, except by a reviewer who may quote brief passages or reproduce illustrations in a review with appropriate credits; nor may any part of this book be reproduced, stored in a retrieval system, or transmitted in any form or by any electronic means, mechanical, photocopying, recording, or other, without written permission from the publisher.

Eighteenth Printing, September 1988

Printed in the United States by Capital City Press

Belanger, Jerome D
Raising milk goats the modern way/Jerry Belanger. — Charlotte, VT.: Garden Way Pub. Co., c1975.

viii, 151 p.: ill.; 22 cm.

Bibliography: p. [149]
Includes index.
ISBN 0-88266-062-4:

1. Goats   I. Title.
SF383.B44   636.3'91'4   75-3493
                            MARC

Library of Congress   76

Contents

1 Basics: What Every Homesteader Should
   Know About Goats   1

2 Milk   9

3 Getting Your Goat   19

4 Housing   33

5 Fencing   43

6 Feeding   45

7 Grooming   69

8 Health   81

9 The Buck   93

10 Breeding   101

11 Kidding   109

12 Raising Kids   117

13 Milking   121

14 Records   131

15 Chevon   137

16 Goat Milk Products   141

   Further Information   149

   Index   151

Introduction

If homesteading, or subsistence farming, had a mascot, it would have to be the goat.

This idea was forcibly brought home to me by a recent book on homesteading in which the author glosses over goats with a single paragraph mentioning that goats seem to be inextricably tied to homesteading and he couldn't figure out why. He didn't care for goats and wasn't going to write about them.

I hadn't given this much thought in the 25 years I've been reading about and studying goats, because they fascinate me. But the comment made me stop to consider why dairy goats are such popular homestead animals.

Historically and globally, goats are providers of household milk as opposed to cows which stock commercial dairies. When commercial dairying becomes feasible, at least in areas that can support cattle, the goat is left behind.

Conversely, when commercial dairying becomes uneconomical or impossible, the goat regains her position as "the foster mother of mankind." (Cowmen use the term without justification since more people in the world drink goat milk than cow milk.) This was true in Europe during both World Wars, it has shown up in the United States during times of economic slowdown, and of course it's even more dramatic in less developed areas.

There is, therefore, some truth in calling the goat "the poor man's cow." Since many people are taking a new look at our planet these days and are deciding there is no inherent value in being "rich" . . . in driving big cars or gorging on corn-fed beef . . . they take what almost amounts to a perverse pride in recognizing the value of dairy goats.

There was a goat boom in the United States during the

1940's. Then interest lagged with growing prosperity, only to return in the 1970's with even more force and vigor. In 1971, membership in the larger of the two dairy goat registries (American Dairy Goat Association) was 1,500. By the beginning of 1974 membership was up to 3,500 and by the end of that year stood at over 5,000! Of course, there are many thousands of others who just "keep goats" or who don't breed purebred goats or for whatever reason have no interest in joining associations.

No one really knows how many dairy goats there are in the U.S. They've been included in the last two farm censuses, but many goats are owned by people who are not farmers and who aren't included in the census. Whatever the number, it's apparent from association membership, registrations, and circulation of goat magazines and interest in goat articles published in other publications, that the number of dairy goats has increased substantially since 1970.

It's true that cow milk is still available in stores, that feed is not rationed (as it was in Europe during the wars) and that in the early 70's even the experts couldn't agree on whether or not we were in a recession. But many people were having a tough time paying grocery bills, and that showed up in increased interest in goats. The goat population could well be as reliable an indicator of the state of the economy as housing starts or auto sales!

However, there are other factors favoring goats today that have never before appeared. The homesteading movement is the focal point. Even though most people are not forced to grow their own food as they might be during a war or depression, they want to. Inflation is one factor. The organic movement is, perhaps, a stronger one: the rapid spread of chemicalized food, including milk and dairy products, and the tasteless, watery store-bought food, including "standardized" milk, have interested people with small plots of land in producing their own.

In addition there is a tremendous need for some form of psychological security to offset the effect of the threat of H-bombs, of impersonal computerized jobs and daily living, congestion, pollution, and the rat race in general.

To top it off, people with time on their hands and money in their pockets can afford to go back to the rural life they or their ancestors never really wanted to leave in the first place. Today they can have jobs in town and the amenities of modern civilization as well as the rewards and virtues of country living.

In this situation keeping a goat is much more sensible and economical than keeping a cow. The physiology of the smaller animal is such that she can produce more milk per pound of body weight on the same amount of feed. The goat is much easier and more economical to house and care for. Many—perhaps most—goats are cared for by women and children for whom cows would be too great a burden and more dangerous. Goats are easier to keep clean.

Of paramount importance is the consideration that several goats can keep a household in milk while a cow would produce far too much for a single family when she's in milk, and none for the two month she's dry. Two or three goats can be bred to freshen at different times and thus maintain a year-round supply. The surplus cow milk, even if fed to livestock or put to other good use, is still a surplus and that means the cost per gallon used in the kitchen increases, especially at today's feed prices. One cow will produce and eat just about as much as six to eight goats.

It's also important to most people that a goat is less expensive to buy than a cow, has a higher rate of reproduction (goats start breeding earlier than cows and usually have multiple births as opposed to the cow's single annual calf) and can be kept on many places where a cow would be out of the question.

But one of the most important reasons for the popularity of goats is an irrational one: goats are simply delightful creatures! After keeping them awhile, most people find goats become family pets as dear as any cat or dog. If they give milk besides, why, any arguments in favor of the cow fade into insignificance.

There is a lot of information on goats. Much of it is hard for the average person to find, or too abstruse and scientific to be of much practical value. Much of it is old and to some degree out of date. A great deal of it centers on goats as show animals, as a commercial enterprise, or relating to situations outside the United States and Canada.

Here we're going to talk about the household goat in North America, animals whose purpose is to provide their owners—modern homestead families—with the best milk and dairy products available anywhere.

Basics:
What Every Homesteader
Should Know About Goats

Before we go into the details of modern goatkeeping let's take a look at some basic facts and terms.

Female goats are called does, or if they're less than a year old, sometimes doelings. Males are bucks. Young goats are kids. They are never "nannies" or "billies." Correct terminology might not be important to some people, but it's mighty important to those who are trying to improve the image of the dairy goat. Most people who think of a "nanny goat" as a mean and smelly beast that produces vile milk will at least have to stop to consider the truth if she's called a doe.

Does are *not* smelly. They do *not* eat tin cans. They are *not* mean! They are dainty, fastidious about what they eat, intelligent (smarter than dogs, scientists tell us), friendly, and a great deal of fun to have around.

Bucks have scent glands located between and just to the rear of the horns or horn nobs, and minor ones in other locations. They smell, but the does think it's great and some goatkeepers don't mind it either. The odor is worst during the breeding season, usually from September to about January. The scent glands can be removed as you will see later, although some authorities frown on the practice, for various reasons.

But bucks have habits that make them less than ideal family pets even if they don't stink. For instance, they urinate all over their front legs and beards, which tends to turn some people off.

But in most cases the homestead dairy won't even have a buck, so we return to the fact that you can keep goats even if you have neighbors or if your barn is close to the house, and no one will be overpowered by aroma.

One of the problems with "goat public relations" is that everybody seems to have had one in the past or knows someone who did. Most of them were pets, and that's where the trouble lies.

A goat is not much bigger than a large dog (average weight is around 150 pounds), it's no harder to handle, and it *does* make a good pet. But a goat is not a dog. People who treat it like one are asking for trouble, and when they get rid of the poor beast in disgust they bring trouble down on all goats and all goat-keepers. If the goat "eats" the clothes off the line or nips off the rose bushes or the pine trees, strips the bark off the young fruit trees or butts people, it's not the goat's fault but the owners'.

Would you let a pig or a cow roam free and then damn the whole species when one got into trouble? A goat is livestock. Would you condemn all dogs because one is vicious . . . after being chained, beaten and teased? Children can have fun playing with goats, but when they "teach" young kids to buck and that kid grows up to be a 200 pound male who still wants to play, there's trouble. Likewise, a mistreated animal of any species isn't likely to have a docile disposition.

The goat (*capra hircus*) is related to the deer: not to dogs, cats, or even cows. It is a browser rather than a grazer, which means it would rather reach up than down for food. The goat also craves variety. Couple all that with its natural curiosity and nothing is safe from at least a trial taste.

Anything hanging, like clothes on the line, is just too much for a goat's natural instincts. Rose bushes and pine trees are high in vitamin C and goats love them. Leaves, branches, and bark of young trees are a natural part of the goat's diet in the wild. If you treat a browsing goat like a carnivorous dog, of course you'll have problems! But don't blame the goat.

Goats eat tin cans? Of course not. But they'll eat paper on tin cans (or any other kind) because paper is made from trees and goats eat trees.

Goats must be penned, for their sake and yours. Each doe

requires roughly 20 square feet of space. They are herd animals, though, and one is likely to be lonesome by herself. Goats do not require pasture, and unless it contains browse they probably won't even utilize it. In any case they'll trample more than they eat. Better to bring their food to them, especially on a land-intensive labor-intensive homestead.

Goats are not lawn mowers. Most of them won't eat grass unless starved to it, and they won't produce milk on it.

Never stake out a goat. There is too much danger of strangulation, and many such goats have been injured or killed by dogs. Even the family pet you thought was a friend of the goat could turn on it.

All this indicates that goats can be raised in a relatively small area, and if there are no restrictive zoning regulations, can be (and are) raised even on average size lots in town. If there are laws against them, it's a good bet that's because somewhere along the line, somebody didn't know how to take care of them. Don't *you* contribute to that situation!

Because goats are livestock, and more specifically dairy animals, they must be treated as such. That means not only proper housing and feed, but strict attention to regularity of care. If you can't or won't want to milk at 12 hour intervals—even if you're under the weather or tired—if you aren't excited by the thought of staying home weekends and vacations and can't count on the help of a neighbor or friend—don't raise goats.

The rewards of goat raising are great and varied, but you don't get rewards without working for them.

The Five Goat Breeds

There are five major breeds of dairy goats in the U.S. Most goats are combinations of two or more of these. (This is not counting Angoras, which are raised primarily in the southwest for mohair and meat, and which probably outnumber dairy goats.) Because there are so many mixtures, it's often impossible to tell which "breed" an animal is.

Nubians are the most popular pure breed, according to the registry associations. They can be almost any color or color pat-

Nubians are readily identified by their pendulous ears and Roman noses. This young Nubian doe was sold for $2,250 in 1974.

tern, but are easily recognized by their long drooping ears and Roman noses. The Nubian is often compared with the Jersey of the cow world. The average Nubian will produce less milk than the average goat of any other breed, but the butterfat content will be higher. Averages can be misleading, though. The Nubian record is 4,420 lbs. of milk and 224.0 lbs. of butterfat in 305 days.

The Nubian is a goat of mixed origin and traces its ancestry to India and Egypt, although the "Anglo-Nubian" now in America was developed in England. The first Nubians arrived in the U.S. in 1909, imported by Dr. R. J. Gregg of Lakeside, California.

The French Alpine originated in the Alps and arrived in the U.S. in 1920, being imported by Dr. C. P. DeLangle. The color of Alpines varies greatly, ranging from white to black, and often with several colors and shades on the same animal. There are commonly recognized color patterns, such as the cou blanc: white neck ("cou blanc" in French) and shoulders shade through silver grey to a glossy black on the hindquarters. There are grey

French Alpines have erect ears and many of them have distinctive color patterns.

or black markings on the head. Another color pattern, the chamoisee, can be tan, red, bay or brown, with black markings on the head, a black stripe down the back, and black stripes on the hind legs. The sundgau is a goat with black and white markings on the face and underneath the body.

Toggenburgs are the oldest registered breed of any animal in the world, with a herd book having been established in Switzerland in the 1600's. They have always been popular in this country, partly because they were the first imported purebreds, arriving here in 1893. Toggs are always some shade of brown with a light or white stripe down each side of the face, white on either side of the tail on the rump, and white on the insides of the legs.

Saanens originated in the Saanen Valley of Switzerland. They are always all white, with erect ears. They are also known for their "dish faces" which are just the opposite of the Roman noses of the Nubians. These are the Holsteins of the goat world since they enjoy the reputation of being the heaviest milkers.

Toggenburgs have white markings on the face and rump.

Saanens are always white and have "dish" or concave faces.

The fifth breed is the LaMancha, a "new" breed of almost earless goat. As with so many new and different things, there is a great deal of controversy over the LaMancha. Some purists register disgust at the thought of such an impure purebred, although they seem to forget that if they go back far enough they'll find that their pure breeds were at one time considered mixtures too. Other people scorn the elfin ears, which admittedly *do* take some getting used to. On the other hand some folks think LaMancha kids look like little Teddy bears, and almost everyone who raises them remarks on their remarkably docile dispositions. In any event, LaMancha backers promote their breed as successfully as any.

As this book goes to press, there are indications that yet another breed will soon be elegible for registration: the African Pygmy. Pygmies are 21 inches tall (or less) at the withers, they vary in color, they are more likely than the other breeds to have triplets or quadruplets, and they are said to be excellent milkers.

The choice of a breed is very much less of a problem than it might first appear, especially for homesteaders. A goat that pro-

LaManchas are noted for their "lack" of ears. Many people claim this is the most docile breed, and many of them are good milk producers.

Pygmies are noted for their small size and white faces.

duces 1,500 lbs. of milk a year is as good as any other goat that produces a like amount, given equality in age, condition, and similar factors. Breed isn't one of them.

Even for those interested in purebred stock the choice isn't made because of any rational factor or breed superiority but because they "like the looks" of whatever breed they choose. Some breeds are also easier to obtain, because popularity of each varies from region to region. You might get a certain doe because she's available, and at the same time make a wise choice because stud service will also be available in your area and there will be a local market for kids of a locally popular breed.

This brief rundown of basic facts should help you decide if you really want to raise goats. I hope you do . . . but with full awareness of what will be expected of *you*. That means you'll want a lot more information on care and management. But before we get to that, let's take a closer look at the product that interested you in goats in the first place: milk.

2

Milk

One of the first questions a prospective goatkeeper who is interested in a family milk supply asks is, "How much milk does a goat give?"

While the question is logical and valid, it's something like asking how many bushels of corn an acre of land can produce. How good is the land, how much fertilizer was applied, what strain of corn was planted, how much of a problem were weeds or insects, was there sufficient heat and moisture? The answer could range from "none" to a quite respectable figure.

To put this in terms that might be better understood by city dwellers, how many ladies' coats can a merchant sell? It depends on whether the seller is in downtown New York or on the edge of small midwestern village; on whether the coats are mink or cloth; whether it's June or December, and so on.

There can be no set answer to the question of how much milk a goat will produce, but here are some considerations.

It must first be understood that all mammals have lactation curves that, in the natural state, match the needs of their young. Man has altered these somewhat through selection to meet his own needs, but they're still there.

The supply of milk normally rises quite rapidly after parturition in response to rapid growth demands of the young. In the goat, the peak is commonly reached about two months after kidding. From the peak, the lactation curve gradually slopes downward.

This brings up what is probably the most common problem with terminology in reference to production. We often hear of a "gallon milker." The term has little or no practical value, for we must know at what point in the lactation curve this gallon day occurred, and even more importantly, what the rest of the curve looks like. The goat that rises to a gallon a day two months after kidding, then drops off drastically and dries up a short time later, will probably produce much less than the animal whose high day is less spectacular but who maintains a fairly high level over a long lactation. Especially on the homestead, where a steady milk supply is required, slow and steady is more desirable than the flashy, one-day wonder.

It's much more practical to speak in pounds of milk per lactation. The traditional lactation period is 305 days. If a goat is to be bred once a year and dried off for two months before kidding for rest and repair, this period is logical. Even though it might be arbitrary in some cases . . . a goat might milk for more or less than 305 days . . . it's a convenient way to compare animals. Cows are judged on the same time span. But it *is* arbitrary, and mainly for record purposes. The homesteader has no need to adhere to such a schedule and in practice even most commercial dairymen milk an animal for shorter or longer periods depending on the production of the goat. Actually, many homesteaders with animals that exhibit long lactations would do well to milk them for two years straight without rebreeding. Production will be lower during the second year, but this might be offset by avoiding a two-month lay-off, breeding expenses, and unwanted kids. It should be pointed out though that not many goats will milk for that long: most will be dry before the ten months are out.

Looking at averages can be misleading, but often that's the only way to get even a rough idea of a situation . . . and it's almost always interesting. There are records of 7,000 pounds of milk in 305 days, and then there are goats that freshen without any milk at all. You obviously won't start out with one of the former and you hope you won't get stuck with one of the latter, but it would be nice to find one that's average.

One of the largest commercial herds in the country posted these annual averages in a recent year.

Breed	Milk (pounds)	Fat (%)	Fat (pounds)
Saanen	1,585	3.5	55
Toggenburg	1,702	3.6	61
French Alpine	1,315	3.5	46
Nubian	1,086	5.1	55
LaMancha	1,459	4.3	63

These are average annual figures for a herd of several hundred animals. They obviously don't include very many zero milkers since this is a commercial herd which would have culled them out. These goats are neither pampered nor neglected. We might say this is an average of "average" goats, not all goats. If it were possible to include all the goats in the country, all the pets, the neglected beasts chained by the neck in weed patches, the low producers kept because the owner is too lazy to cull or because he's waiting for some sucker to come along so he can turn a profit on a poor investment . . . the "average" production would be much lower. These figures given are for decent, well-cared-for stock.

However, this is just for one herd. Another large herd has an average of 1,700 pounds. Still another, smaller herd (40 animals) has racked up an average of 2,600 pounds. But since all these figures come from herds of established, experienced breeders who know their business and have culled heavily, they're submitted for your admiration, not envy. If a beginner takes home even a top doe, production will undoubtedly drop.

There is some indication that the average production per goat has declined somewhat over the years. In 1945, for instance, the average production of goats on test was as follows:

Breed	Milk (pounds)
Saanen	2,325
French Alpine	1,895
Toggenburg	1,902
Nubian	1,626

It isn't really statistically valid to compare these averages with those given above because the former were from one herd. However, data from England *does* show the same decline, and some people blame the changes in feeding methods.

Record-breakers are even more meaningless for the homesteader than are averages. The new goat owner has about as much chance of buying a fantastic producer as the guitar pickin' kid down the block has of coming up with a hit song on his first record. It takes knowledge, experience and work to come up with a winner in any field. But at least the figures will show you what goats are *capable* of.

The French Alpine record of 4,826 lbs. was set in 1968; the top doe for 1972 (latest records available) produced 4,120 lbs. The Nubian record was set in 1972 (4,420 lbs.) but the runner-up produced 3,650 lbs., a lot less. The U.S. Saanen record was set in 1971 with a lactation of 5,496 lbs., but the top Saanen in 1972 produced 4,160 lbs. and the runner-up gave 3,920. (The world record was set by a Saanen in Australia in 1972: 7,546 lbs. of milk with 223 lbs. butterfat. Much more effort is expended on upgrading dairy goats in Australia than in the U.S., including government research which is unknown here.)

The Toggenburg breed leader produced 5,750 lbs. in 1960. The leader in 1972 produced 4,250 lbs.

Figure 1 shows some actual production records of a small herd of Nubians. The top doe in the herd produced 2,150 lbs. in 10 months, the bottom doe 1,300 in nine months, and the herd average was 1,730 lbs. Notice the lactation curve. The average production goes from 7½ lbs. at kidding to about 8 lbs. two months later. From there it tapers off to about three pounds 10 months after kidding.

But even this hides some of the reality. Figure 2, while harder to follow, paints a more accurate picture.

These are actual, individual records from a herd of four does. It shows how much production can vary among animals. One doe had a 17 month lactation. She gave 1,800 lbs. in the first 10 months and continued to produce a steady 5-5½ lbs. daily until pregnancy caused production to drop. Another doe reached her peak in the fourth month.

If you owned these four does and were going to sell one,

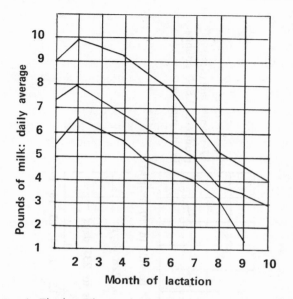

Figure 1. The best doe in this herd produced 2,150 pounds of milk in 10 months. The lowest record shown is 1,300 pounds in nine months. Average for the entire herd was 1,730 pounds in ten months.

which one would it be? Remember that when you buy your first goat.

Note that one doe produced more than twice as much as another even though they ate about the same amount of feed and required the same amount of care. Note also that two weren't worth milking after only eight months.

Age is a factor in milk production. Records from this same herd show that peak production comes in the fourth or fifth year. But there are also other factors involved, and any individual goat can vary erratically from one year to the next. Even on a day-to-day basis milk production is affected by changes in weather, feed, sickness or injury, outside disturbances, and other factors.

The question "How much milk does a goat give?" can only be answered by another question: "How long is a piece of string?"

EAU CLAIRE DISTRICT LIBRARY

Figure 2

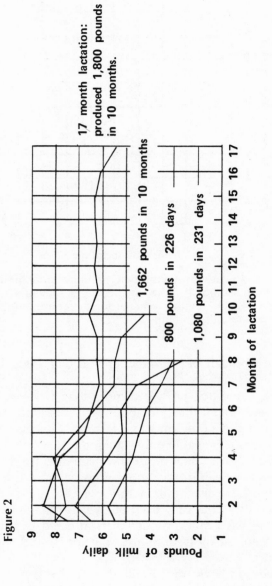

14

From the foregoing you can assume that you'll be able to find a goat that will produce a respectable amount of milk for your table. Next question: will your family drink it?

When you've been raising goats awhile you'll surely be asked what you do with the milk. Many people seem to assume goat milk is used only in hospitals. The healthful aspect of goat milk is as legendary as their aroma and their preference for tin cans . . . and perhaps just about as detrimental.

Many doctors do, in fact, prescribe goat milk. Many more would if steady supplies of sanitary milk were available. It may be recommended in cases of dyspepsia, peptic ulcer, and pyloric stenosis. It is preferable to cow milk in many cases of liver dysfunction, jaundice, and biliary problems because the fat globules are smaller (2 microns versus $2^1/_2$ to $3^1/_2$ microns for cow milk). Goat milk has been used for infants being weaned, children with a liability to fat intolerance or acidosis, infants with eczema, pregnant women troubled by vomiting or dyspepsia, and nervous or aged people with dyspepsia and insomnia.

Goat milk is more easily digested than cow milk because the fat is finer and more easily assimilated; it is particularly rich in antibodies; and when freshly drawn has a much lower bacterial count than cow milk. When whipped, cream from goat milk is bulkier than cow cream: the specific gravity of cow milk is 0.96 and of goat milk 0.83.

But despite all that, goat milk is not medicine! It's food. Good food. It is drunk by more people in the world (or used in cheese or yogurt) than cow milk.

Because most Americans aren't familiar with the product they have many misconceptions about it—misconceptions based on the comic strip image of the goat, or on the unfortunate experiences of a few who have been exposed to goat milk produced under conditions that make it unfit for human consumption.

The home wine-maker who takes a basket of over-ripe and spoiled, wormy, moldy fruit, puts it in a dirty crock and pays no attention to proper fermentation, does not end up with a fine wine. And the goatkeeper who milks a sickly, undernourished animal of questionable breeding into a dirty pail, lets the milk "cool" in the shade and serves it in a filthy cup does not end up with fine milk. And while no one would disparage all wines after tasting the concoction just mentioned, many people are all too willing to write off all goat milk after one experience.

Milk is almost as delicate a product as fine wine. It must be handled with knowledge, and care, whether it comes from a cow, goat, or camel.

Goat milk does not taste different than cow milk. It doesn't look appreciably different. It is not richer. It certainly does not smell!

Most goat raisers enjoy serving products of their home dairy to skeptical friends and neighbors. The reaction is invariably, "Why, it tastes just like cow milk!" (I have noticed, however, that city people who are accustomed to regular standardized milk—milk which has butterfat removed to just barely meet minimum requirements—or worse yet who drink skim milk, are prone to comment on the "richness" of goat milk. They'd say the same thing about real cow milk if they had the opportunity to taste it before the technologists started messing around with it.)

In rare cases there *are* animals that give off-flavored milk. We can't call it "goaty" because some cows have the same problem. It can be caused by strong flavored feeds, by mastitis, and in rare cases by genetic factors. The latter is a hopeless case and the animal should be culled. It's not a bad idea to taste the milk of a doe you're buying, if possible.

Many people assume the milk of goats is richer than cow milk. Standardization of cow milk just mentioned is one explanation for this belief. The percentage of butterfat varies with breed, stage of lactation, feed and age . . . with goats as well as cows. On the average there is virtually no difference between the two.

It has long been said that goat milk is "naturally homogenized" because of the smaller fat globules. Natural homogenization might be a problem for homesteaders who want cream for butter and other uses: goat milk requires a separator to extract the cream. Actually, it probably isn't the size of the fat globule that causes the cream in goat milk to remain in suspension. Research has shown that goat milk lacks a fat-agglutinating protein, an euglobulin. Actually the cow is probably the only domestic animal which produces milk with this particular protein, according to Prof. Robert Jenness of the University of Minnesota. Sow and buffalo milk do not form creamlines. Because homogenized cow milk is the norm in this country, most people would probably be surprised to see cream rise in milk! They won't be shocked by goat milk.

What else can we say? At the very least, your home-produced goat milk will be every bit as palatable and healthful as the milk you now buy in plastic cartons. If your kids are like mine, they'll soon be holding their noses to drink cow milk!

3

Getting Your Goat

So you've decided to buy a goat. Now the problem is finding one, and perhaps more importantly, finding the right one.

This isn't always easy. As recently as 1967 there were fewer than 5,000 goats registered with the American Dairy Goat Association (ADGA), the largest goat registry in the U.S. This figure crept up slowly until 1971 when it jumped 1,300 in a single year. Registrations increased more than 3,100 in 1972. And it's expected that the total for 1974 will top 20,000!

These are only registered animals: no one knows the **real** total. And they are fairly concentrated, mostly on the west coast, and in pockets of the midwest and eastern dairying regions.

If 20,000 goats sounds like a lot, notice that there are 3,000,000 cows . . . in Wisconsin alone.

In terms of raw numbers it might appear that it's five times easier to find a goat today than it was a few years ago. But since many more people are looking for goats, it's still pretty much a seller's market.

If you have neighbors who have goats or if you have seen goats in your area, you have a good lead. Most goat people, especially serious ones, know who the other goat people in their area are. If they can direct you to a local club many of your troubles are over.

You might be able to track down a goat at a state or county fair. Goat shows are becoming more common at these events, although at many of them, goats are still treated like "the poor man's cow."

The ideal place to see goats, learn about goats and talk about goats is at one of the better fairs or at a goat show. The best of these will have some of the top goats in the nation on display, which means some of the best and most knowledgeable breeders to talk with.

Goats are sometimes advertised in the classified sections of newspapers and farm periodicals. Also, be sure to check the ads in *Dairy Goat Journal,* Box 1908, Scottsdale, Arizona, and *Countryside & Small Stock Journal,* Waterloo, Wisconsin 53594. Look for breeders near you who you can visit. It is possible to ship goats, but it's not a good idea for your first one. You'll need to learn more about them, and of course you want a first-hand view of pens, mangers, fences, other equipment . . . and goats.

If you're like most people who just want a family milker, you'll end up buying what's available regardless of breed, type, conformation and desirable traits. But of course you'll still want to have some idea of what to look for.

We've discussed breeds. In each breed you will also find these classifications: registered purebred; unregistered purebred; American; registered grade; and grade.

A purebred animal is one with a pedigree that can be traced, through a registry association's herdbook for the breed, to the very beginnings of the recognition of the breed as "pure". If the purebred is registered, its offspring will also be able to be traced. If it is not registered, it will be all but impossible to prove that the offspring, or even the animal itself, is purebred. Likewise unless there are registration papers on the sire and dam it will be impossible to register the animal or to prove that it's purebred.

An animal without a pedigree is considered a grade. It may be pure, but without proof of who the parents were, you can't prove it. A grade more commonly is a mixture of two or more breeds.

A pedigree is merely a paper showing the ancestry of an individual animal. Registration papers are official documents showing that the goat is entered in the herd book of a registry association.

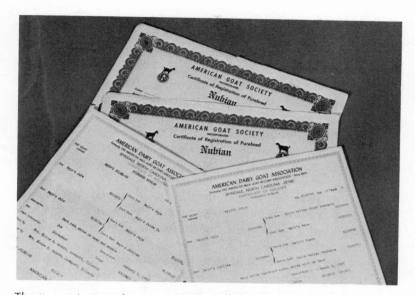

There are two registry associations that maintain herd books on dairy goats in the United States. Registration certificates are not licenses to milk; they are merely proof that the goat is entered in the herd book of a registry association.

Because some grades are very good animals, the ADGA has a program of "recorded grades." The animal must meet certain requirements and is recorded as a grade of whichever breed it most closely resembles.

If such a recorded grade is bred to a registered purebred buck, the offspring will be one-half "pure." Only does of these matings are recorded. If such a doe is bred to a purebred, her kids will be 3/4 purebred and in one more generation the result will be a goat that is 7/8ths "pure". It was considered that further upgrading would make little difference so 7/8ths was taken as an arbitrary cutoff point for a special category called "American". (The ADGA board of directors voted to change this to demand one more generation, or 15/16ths pure, at the 1974 annual meeting. There was considerable debate on the topic and the general membership meeting set aside the ruling temporarily, but chances are by the time you read this "American" will mean 15/16ths pure.)

Registration papers don't mean anything more than that the goat is listed with one of the registry associations and that the pedigree can be traced back to the closing of the herd book. This can be important to people who are familiar with and looking for certain blood lines. For the homesteader interested in milk and knowing nothing about goats' family trees, the papers don't mean much.

Don't reject purebreds out of hand, though, before you read the chapter on record keeping. While purebreds might or might not give you more milk, they could very well reduce the cost of your milk, as that chapter explains.

Only registered goats can be shown in official ADGA and AGS shows. AGS does not register grades or Americans, but has been considering opening a herdbook for LaManchas (which are Americans under the ADGA system).

The homesteader should be more careful about buying purebred goats than grades. Too many poor goats have been kept, and used for breeding, simply because they had registration papers. In some cases the animals had good type but little milk and were kept for show purposes because of that type. In other cases they didn't even have that much going for them.

This obviously doesn't mean that a grade is inherently better than a purebred: there are good and poor specimens in each category, but the purebred will generally cost more, making mistakes more expensive. So how do you decide what goat to get?

It takes a great deal of experience and study to be able to tell a "good" goat from a "poor" one. To the homesteader, a good goat is simply one that milks well for a reasonably long lactation. But since you can't know for sure what she'll put in the milk pail until you get her home, it helps to have some idea of what she needs to do the job.

The general appearance of livestock, the way it's put together, is called conformation. These are some of the points of conformation goat breeders look for:

The head should be moderately long with a concave or straight bridge to the nose, except in the Nubian which must have a definite Roman nose. Saanens have a concave nose, or "dish face."

Ears are a part of conformation that hold little interest for the person seeking household milk. As LaMancha breeders say, you don't milk the ears. But to make this description more complete, note that the ears should be pointing forward and carried above the horizontal, with the exception once again of the Nubian, which must have a long, thin-skinned ear, hanging down and lying flat to the head. LaManchas, the "earless" breed, have a size limit of two inches on their ears. The airplane ears that result from a cross between a Nubian and another breed are ridiculed by many, but again, you don't milk the ears and more than a few homesteaders have such animals and love them.

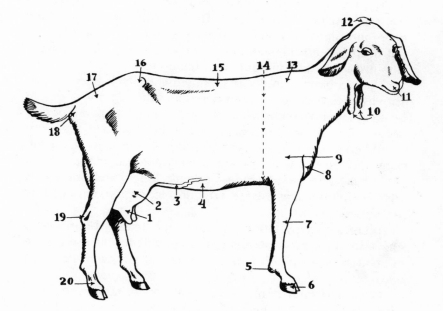

Diagram Showing Parts of a Goat's Body:

1. teats	6. hoof	11. muzzle	16. hip bone
2. udder	7. knee	12. knobs	17. rump
3. milk vein	8. chest	13. withers	18. pin bone
4. belly	9. shoulder	14. heart girth	19. hock
5. claw	10. wattles	15. back	20. pastern

Of more importance to the home milk supply are such points of conformation as the muzzle, which must be broad with muscular lips and strong jaws, as this is an indication of feeding ability. Large, well-distended nostrils are essential for proper breathing.

The neck should be clean-cut and feminine in the doe, masculine in the buck, with a length appropriate to the size of the animal. It must blend into the shoulders smoothly and join at the withers with no "ewe neck". The goat needs a large, well-developed windpipe.

The forelegs must be set squarely to support the body and well apart to give room to the chest.

The rib cage should be well sprung out from the spine with wide spacing between each rib. The chest should be broad and deep, indicating a strong respiratory system. The back should not drop behind the shoulders, but should be nearly straight with just a slight rise in front of the hip bones.

The hip bones should be slightly higher than the shoulder. The distance between the hip bones and the pin bones should be great, but not so long as to make the animal look out of proportion. The slope of the rump should be slight and the rump should be broad. The broader the rump, the stronger the likelihood that the goat will have a high, well-attached udder.

The barrel should be large in depth, length and breadth. A large barrel indicates a large, well developed rumen necessary for top production.

There are many types of udders and teats. Abnormalities such as double teats, spur teats, or teats with double orifices are to be avoided. (On very young kids, extra teats can be clipped off.)

While very large "sausage teats" are undesirable, very small ones may be worse as they make milking difficult, especially for people with big hands, and it takes longer. However, many first fresheners have tiny teats which quickly become more "normal" with milking.

Don't be impressed by large udders, as many of them are just meat. With a very pendulous udder you'll have to milk into a pie pan because there isn't room to get a pail under the goat! Stay away from those, if possible. Of even more serious concern,

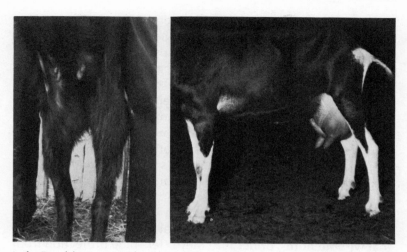

Left: Double teats such as these make a goat useless as a milker. Right: A fine example of a goat udder. There are many variations in shape and size of teats and udders, but a large udder doesn't necessarily mean that the goat gives a lot of milk.

pendulous udders are more prone to injury and mastitis infections.

A well attached udder, carried high out of harm's way, with average size teats and free from lumps and other deformities, is the heart of your homestead dairy.

The condition of the skin reflects the general condition of the entire animal. It should be thin and soft, and loose over the barrel and around the ribs. A goat with "unhealthy looking" skin probably isn't too healthy. Check for mites and lice.

If you're looking at younger stock, avoid the overdeveloped kid. A kid that develops too early seldom ends up being as good an animal as one that has long, clean lines and enough curves to indicate that the framework will be filled out at the appropriate time.

Many goats have wattles, which are small appendages of skin usually found on the neck, although they can be just about anywhere. They are a family trait, not a breed characteristic: some animals of all breeds have them, others don't. They are merely ornaments. Some breeders cut them off from young kids

not only to make the animal look smoother but because sometimes another kid will suck on them and cause them to become
sore. Cut with a sharp scissors at an early age they will bleed but
little, or you can tie a thread tightly around the base and let
them fall off.

Horns, likewise, are indicative of neither sex nor breed. Some
animals have them, some don't. It's common practice to remove
them at birth because they can cause problems later. Some people claim horns are dangerous to both other goats and their
keeper, and there's little doubt that it's tough to build a manger
that will accomodate a good set of horns. These ornaments are
a disqualification for show goats.

On the other hand, some people like horns because they
think they're beautiful, because a goat just doesn't look like a
goat without horns, or because they offer some protection from
dogs and other enemies. The dogs will probably win anyway,
but horns do tend to even up the odds.

Disbudding young kids is much easier and safer than dehorning older goats. See the chapter on care and grooming for
more details on this.

If a good goat to you is one that milks well, first consider
overall appearance and conformation, because a sickly animal
or one that isn't built like a dairy animal isn't likely to do the
job for you. Then consider any records that might be available.

Show wins can be impressive (and they tell you what quali-
fied judges think of the animal's conformation) but, like registration papers, blue ribbons are no license to fill a milk pail.

There are, however, records that mean something for people
who just want milk. The ones you'll most likely run into are barn
records, the daily tally sheets kept by the owner which show
each milker's production.

To accept these barn records at face value you'll first have to
judge the character of the owner, which might be more difficult
than judging the value of the goat. In fact, even without records
of any kind your best assurance of getting a fair shake is by buying from someone you feel you can trust. Such a person will
help you learn and some will even stand behind the animals
they sell . . . although this is asking a lot, sometimes, as when
careless or ignorant people take home a good goat, neglect or

abuse it, then complain that they were ripped off because the animal doesn't meet their expectations. There are also renegades who will sell you "registered" goats with the papers to come later (they never do) or who are more interested in disposing of a goat or acquiring your cash than they are in helping you or promoting goats. Almost as bad are the people who have been raising goats for years—and who still don't know as much as you will after reading this book!

People who raise goats, in other words, are a cross section of the general population. If you buy a goat without knowing very much about goats, it will help to know something about people.

The barn records will be in pounds and tenths. A quart of milk weighs, for all practical purposes, two pounds. Eight pounds, then, is a gallon. Be wary of milk records expressed in pints and quarts (and downright skeptical of milk expressed in gallons!). Even an honest and well-meaning milkmaid can be misled by a bucket of foaming milk that "looks" like three pints. Weight is much more reliable.

Barn records depend entirely on the accuracy of the scales and the integrity of the milker. They can be falsified or altered. So there are records that mean more. Unfortunately they aren't widely used by goat owners because of cost and other factors, but their use is becoming more widespread.

Cornell University was computer processing Dairy Herd Improvement Association (DHIA) records for 423 herds as of October, 1974: an increase of 161 in just one year. The average herd had five milkers.

DHIA is a system whereby testers visit dairy herds on a surprise basis to weigh milk and test it for butterfat. Testing was customarily done one day a month, on a morning and night milking, but with computerization it has been found accurate at even longer intervals. Since the cost is too high for many small goat herd owners ($15-20 a month) DHIA herds are still relatively rare.

Another term you'll find is Advanced Registry, or AR. An AR doe is one that has given a certain amount of milk in a year. The amount varies with the age of the does and other factors, but the AR designates the doe as a good milker.

Still another term referring to production is star milker, or "* milker." Unlike the AR, the star is based on a one-day test rather than on an entire lactation. Like AR, points are based on the stage of lactation the doe is in. It should be noted that many does who have earned Advanced Registry certificates cannot become star milkers because they never give enough in one day. Even though weighted for the length of lactation, a doe still has to produce around 10-11 pounds in a day to earn a star.

Conversely, many star milkers wouldn't be able to earn their AR because they don't produce enough in the entire 305 day lactation to qualify. Both are official, however, and both are supervised by someone other than the owner.

(If you would like to have your goats on test, contact ADGA for details.)

The star is based on total pounds of milk produced in 24 hours; the total number of days in the current lactation, based on 0.1 percent for each 10 days with a three point maximum; and the butterfat percentage. A doe that gave 6.4 lbs. in the morning and 7.0 lbs. in the evening would earn 13.4 points. If the butterfat was 3.7 percent, 3.7 times 13.4 equals 0.4958 lbs. But since butterfat is divided by 0.05 for point purposes, this amounts to 9.91 points. If the doe was fresh 44 days this would add 0.44 points for lactation. The total would be 23.75 points. A minimum of 18 are needed to earn the star.

If a doe's dam has earned a star and she earns one herself, she becomes a "** milker." If her granddam also had a star, she is a "*** milker."

Bucks can also have stars, signifying those earned by their maternal ancestors.

With or without any of these papers . . . registration, pedigree, show wins, barn or official test records, advanced registry certificates or stars . . . many people recommend that prospective goat buyers see the goat being milked, or better yet milk her themselves. For an inexperienced milker the doe will naturally be nervous, and probably won't produce much. But it's better to get the practice and let the owner finish the job than to get her home and find out you can't milk. Taste the milk to check it for off-flavors that might result from disease or a genetic conditions. (It might just be something she ate, too.)

Price of Goats

The cost of goats probably generates more heat among goat raising homesteaders than anything else. Some plug for higher prices, some for lower, and there are good arguments on both sides.

Nationwide, the average price for a registered doe kid in the spring of 1972, according to the goat editor of *Countryside & Small Stock Journal,* was $75. A well-bred buck kid out of an exceptional dam was selling for $100-$150. According to that survey, LaManchas had higher price tags than other breeds. Milking does went for $300-$400.

If that seems high, consider that the top price paid for a goat in the 1971 ADGA Spotlight Sale (an auction of the best animals for sale in the nation) was $1,800. In 1972 the highest bid was $1,700; in 1973 $2,100 and 1974, $2,250.

It's perfectly true that in many localities you might be able to pick up a goat for ten or 15 dollars, and sometimes for nothing. From the buyer's standpoint, there's certainly nothing wrong with that, providing the goat suits his needs. Too often, though, you get just what you pay for. Some people want to get rid of a goat because it was just a pet the kids got tired of, and others want to dispose of worthless stock for whatever they can get. In some cases fairly good goats go cheap simply because the owner doesn't know what goats are worth.

It cost $70 or more to raise a kid from birth to one year of age even before feed prices skyrocketed. There's no law saying somebody can't sell that goat for $25 or even $10. But anybody who sells goats that way has a pretty expensive hobby. (If they're selling all their stock there must be a reason: make sure the reason is not that the goats don't produce any milk!) A homesteader may not be in business, but everything is supposed to pay its own way and the homesteader who sells goats for less than they cost is in effect raising the cost of his own homestead-produced milk.

There are large, well-established markets for hogs, cattle and other livestock, yet the prices fluctuate wildly.

There are no such markets for goats, so it's even more difficult to establish a market value at any given time. One method is to figure that a cow gives as much milk as 6 to 8 goats, so a goat should logically be worth 1/6-1/8 of the going price of cows in your area. A somewhat better, though more difficult method, is to determine how much a goat is worth to your homestead using the information found in the chapter on record keeping.

Most homesteaders looking for a goat want to buy one that will produce milk for them now, today. At the same time they wonder if they could save money by buying a kid which would cost less than a mature milker.

The kid, by the time she's grown, bred, and fresh, will cost as much or more than the milking doe, and you will have gone for a year or more without milk.

The milking doe, on the other hand, will almost certainly drop in production when you move her to a new home, and if you're an inexpert milker or it's late in her lactation, she would very well dry off. So you still won't have any milk.

Probably one of the best ways to learn and still face only a minimum delay in producing your own milk is to buy a bred doe in the fall. You might get some milk from her, and at least you won't be an amateur when she kids in the spring. You'll get to know each other before she kids and she'll be at home when she freshens. One problem with this approach is that goats are more scarce, and more expensive, in the fall, because of the problem of getting enough winter milk. Any good doe that is still milking in late fall isn't likely to be sold.

Goats are more plentiful in the spring when barns are bulging at the seams with newborn kids most of them weren't designed to hold.

An older goat might be a bargain if you acknowledge beforehand that you're getting her to gain experience and that you might lose her sooner than you'd like. On the average, you can expect a goat to produce for 7-10 years.

All of this probably makes buying a goat sound frightfully complicated. It isn't. You'll probably mentally file some of this information then promptly forget most of it. That's all right. At least you'll have some idea of what to look for, and chances are excellent that you'll want to refer back to this information later when you start to upgrade or look for more goats.

In the meantime, buy a goat. It's the best way to learn. But before you bring it home, read the next chapter and be sure you have the proper housing and equipment ready.

4

Housing

Many people, especially those who haven't had much experience with livestock, are prone to bring home an animal and *then* decide where and how they're going to keep it. This is definitely starting off on the wrong foot.

Most people contemplating raising goats already have facilities that, with a little work, will serve as shelter. (Anybody new to goats would be well-advised to learn something about them before building brand-new facilities: a few years' experience will go far toward eliminating costly mistakes.)

Goats do not have to be kept warm even in northern climates if they've been conditioned to the cold through the fall. But in any climate, their housing must be dry and free from drafts. Goats are very susceptible to pneumonia.

Goats are commonly kept in garages and sheds, old chicken coops and barns. These may be wood, concrete, cement block or stone. They may have wooden floors or the floors might be dirt or concrete. All of these are acceptable.

Ideally, the goat house should be light and airy with a southern exposure. It should be convenient to work in, which means the aisles and doorways should be wide enough to get a wheelbarrow through without barking your knuckles; feed and bedding storage should be conveniently nearby; and running water and electricity should be available to make your work easier, more pleasant and safer.

Wooden floors such as found in brooder houses or other poultry buildings can be warm and dry if the rest of the structure is snug and tight. But wood rots, so special precautions are needed. The bedding must be highly absorbent, and changed frequently. The very best bedding for any type of floor is peat moss. This material can absorb 1,000 lbs. of water for each 100 lbs, of its dry weight . . . far more than any other bedding material. But if you have to buy it, it's expensive, which means it isn't very widely used.

Chopped oat straw rates second among commonly used materials, but it only absorbs 375 lbs. of water per 100 pounds of dry weight, less than half of what peat moss can handle. Note too that this is *chopped* straw: long straw will only absorb 280 lbs. per cwt. Wheat straw is somewhat less absorbent than oat straw.

Wooden floors are obviously not the most desirable for animals like goats, where large quantities of wet bedding will accumulate. You wouldn't put wooden floors in a new building

Absorbtive Bedding Capacities of Various Materials

Material	Pounds of water absorbed per cwt dry bedding
Peat moss	1,000
Chopped oat straw	375
Oat straw, long	280
Vermiculite	350
Wood chips (pine)	300
Wood chips (hardwood)	150
Wheat straw, chopped	295
Wheat straw, long	220
Sawdust (pine)	250
Sawdust (hardwood)	150
Corn stalks, shredded	250
Peanut hulls	250
Sugar cane bagasse	220
Corncobs, ground	210
Broadleaf leaves	200
Sand	25

designed for goats. But if you have the building there's no reason why you couldn't use it.

Concrete floors are only somewhat less desirable, according to many experienced goat raisers. Concrete is cold. The urine cannot run off (which is fine with most homesteaders, who want to accumulate it for organic gardening anyway). That means it takes a great deal of bedding. But this is no great drawback in most cases, especially given the nature of the goat.

Cow manure is extremely loose and liquid, certainly in comparison with nanny berries. Those neat little compact balls bounce, and in this context bouncing is to be preferred over splashing. With a little fresh bedding to keep the top surface clean, goat litter can accumulate to a considerable depth and still be much less offensive than a cow stall. The problem here, with concrete floors, is that while the surface may be quite clean and dry, the bottom layers can be swampy morass of maggot breeding ground. Deep litter does not signify a sloppy farmer: for goats, the deeper the better, certainly on concrete floors and especially in winter. The lower layers will actually compost in the barn, not only helping to warm the goats' beds but hastening its use in the garden. Some people use compost activator on the bedding to speed up the bacterial action. Such deep litter is warm, and quite odorless. (Until you clean the barn, at least.)

Concrete floors are easy to get really clean, which can be very important in the summer when deep litter may not be so desirable. Goats prefer less bedding in the warmer months, but even then concrete floors require special management. Concrete is always relatively cool, and hard, and there have been reports of rheumatism-like problems arising among goats on hard surfaces. Sleeping benches should be considered on concrete areas, for both winter and summer. There have been cases where people have torn out concrete floors to install what they consider the ideal: dirt or gravel floors.

Earthen floors are certainly the easiest to maintain. The urine soaks away, and much less bedding is needed. Earth is warmer than concrete, and more comfortable for the animals. But that loss of urine can be of concern to the organic homesteader. They'd rather use more bedding, which is important for soil improvement too, than to let those nutrients leach away.

Everything depends on your unique situation. Some years ago there was a lady who kept goats in what was practically the center of town, as the suburbs grew up around her. She kept the animals on concrete floors, without bedding. The urine drained away, and the droppings were swept up daily. The place was spotless, there was never a complaint from the neighbors, and the only waste disposal was a daily coffee can of nanny berries that went on the rose bushes.

The floor and its proper maintenance will contribute a great deal to the health and comfort of your goats.

It is assumed that the roof will not leak and that drafts are prevented from entering the house through cracks and around windows or doors. Insulation can be highly desirable in most areas, but special precautions will have to be taken to protect it from the goats. Most wall materials will be chewed, eaten, or smashed. Plywood, plasterboard and the like won't last more than a matter of days. If you have access to stout planks, fine. Cement wallboard can also be used, and will make a cleaner wall.

Most goats today are raised in what is called "loose housing." That is, instead of being chained in individual box stalls or stanchioned like miniature cows, they are free to move about in a common pen. While many cow dairies are converting to this system, it makes even more sense for goats. Goats are herd animals. They need companionship. They are active animals and need exercise. And loose housing is a whale of a lot less work than individual box stalls or tie stalls. Loose housing obviously entails lower original cost in both time and material. It's more flexible: if you have four stanchions there is no way you can house five goats.

The size of the building is dependent on several factors. Recommendations range from 12 to 20 square feet per animal (less in warm climates where they spend more time outside). If there is a sizeable pasture or exercise yard and the barn is used mainly as a dormitory, you could get by with a smaller figure. But even that is only one consideration.

The homestead goat raiser will need space for hay, bedding and grain, and more likely than not the milking will be done in the barn too. If you're working with minimum space requirements, don't forget that the addition of kids will require more space. And then too, there is always the possibility that your herd will increase . . . goat herds have a way of doing that, even on well-managed homesteads.

Here again, almost any situation can be utilized. If there simply isn't space for hay in the building you have in mind, it can be stored elsewhere. More labor will be involved, but it won't keep you from keeping goats. Grain can be stored in garbage cans in other outbuildings if necessary. Milking elsewhere can be extremely inconvenient, however, especially in inclement weather. Unless an alternate shelter is very close to the goat shed, try to incorporate milking space into the main building.

Grade A dairies, cow or goat, *must* have a separate room for handling milk. Where large numbers of animals are handled, a separate milking parlor is advisable to eliminate dust and barn odors. But for the homesteader with a few goats, kept clean, milking in an aisle is quite acceptable and far more convenient. A milking bench may not be a necessity, but it will contribute greatly to ease of milking, and will result in a superior product.

The major item of equipment in a goat house is the manger. These can be constructed in many ways, a few of which are pictured here. Goats are notorious wasters of hay, and this is the main factor to consider in designing a good manger. Grain is often fed in mangers, since most goats won't get their allotted ration during milking. Greed can cause problems then, unless some means of fastening the animals into the manger is devised to prevent bossy does from taking more than their share. Hay can be fed free choice in the same manger: the goats can come and go as they please, and eat as much of it as they want.

There is much variation in jumping ability, or perhaps desire. Contented goats are less likely to leap fences of any height. But if a deep litter system is going to be used, remember that a four foot fence in October may be only three feet high a short time later! (This same observation applies to ceiling height, naturally.)

Gates and latches are important in goat houses. They should be sturdy, for goats love to stand on things with their front feet, and gates are the favored place in any fence to do this standing. With deep litter, the gate should swing out of the pen, which is a good idea in most cases anyway. Make sure the gate is wide

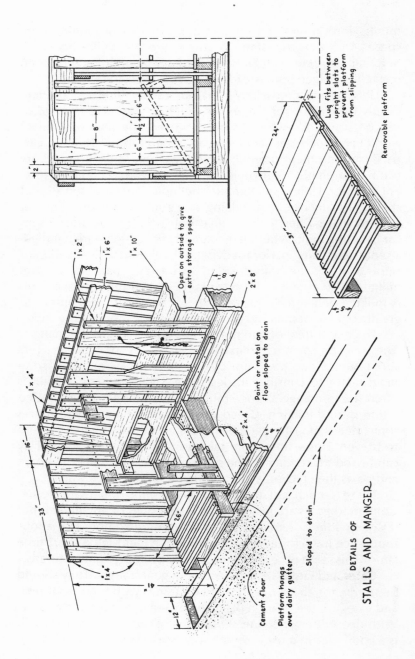

DETAILS OF
STALLS AND MANGER

Lug fits between upright slats to prevent platform from slipping

Removable platform

24"

37"

5"

Open on outside to give extra storage space

1" x 2"

1" x 6"

1" x 10"

8"

2" x 8"

Paint or metal on floor sloped to drain

2" x 4"

4"

1" x 4"

16"

33"

26"

1" x 4"

4"

Cement floor

Platform hangs over dairy gutter

Sloped to drain

12"

8"

6"

4½"

6"

8"

2"

38

enough to get through with a wheelbarrow or whatever you'll be using at cleaning time. Since most goats are Houdinis when it comes to unhooking latches, pay special attention to those items.

While some homesteaders demand more luxuries and conveniences than others, two are of particular interest: water, and electricity. Water piped to the barn can save countless minutes, which on an annual basis amounts to hours. The goats are more likely to have a continuous supply of fresh water if you don't

A manger is necessary for feeding hay and grain. Goats are allowed to eat as much hay as they want, but the grain is rationed: usually one pound per day for maintenance and one additional pound for each two pounds of milk produced.

have to lug it a long distance, and from that standpoint alone can be worthwhile.

Electricity is obviously a boon when you have to do chores before or after the sun shines. Trying to milk by flashlight is hectic, and lanterns can be dangerous as well as troublesome. Electricity, moreover, will sooner or later be wanted for clippers, disbudding irons, and other possibilities.

You'll want storage space—how much depends on the type of operation you have. There should at least be room for a pitchfork out of harm's way; hair clippers; hoof trimming tools; brushes; disbudding iron or caustic; and perhaps a medicine cabinet. Provide a place for hanging scale, and make sure the milk records are kept where the goats can't nibble on them!

Milking equipment will be stored somewhere cleaner than the barn, of course.

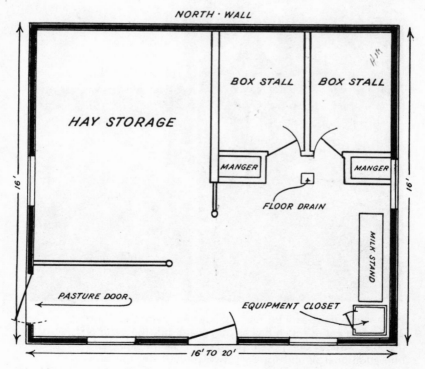

Plan for a two-goat barn

Buildings should be whitewashed inside, or painted with a lead-free white paint. This will make the building not only more attractive and pleasant to work in, for you and the goats, but it will be cleaner, and light colors tend to discourage flies and other pests.

Most homestead dairies use the kitchen as a milkhouse. The ideal milkhouse—well ventilated, with hot and cold running water, rinse sinks, floor drain, impervious walls and ceiling— is nice, but a bit much to expect for a dairy with only a couple of goats. The kitchen works just fine for most people. That's where the utensils are washed and kept.

This chapter includes some suggested floor plans. They will have to be adapted to your particular circumstances. After a few months of doing chores, you'll probably want to make some changes based on your building, your animals, and the way you do things.

Facilities for dairy goats need not be elaborate or expensive. But because they are *dairy* animals, you want to keep them and their surrounding as clean as possible, and because nobody does unnecessary work just for the fun of it and because you want the best production your does are capable of, you'll want to plan quarters that are easy to keep clean; are pleasant for both you and the goats to be in; and will contribute to the health and well-being of your herd.

5

Fencing

Fencing is probably more important—and more difficult—with goats than with any other domestic animal. Goats will jump over, crawl under, stand and/or lean against, and in any other way they can think of circumvent any boundary that is not strictly goat-proof.

For most people a lot of fencing really isn't necessary. Don't think in terms of pasturing your goats, at least at first, and especially if you don't have browse—trees and shrubs and brush, rather than grass. Goats that are properly fed in the barn will probably just ignore even the finest pasture . . . although they'd be delighted to get at your prize roses!

In this case a small, sunny, exercise yard is sufficient.

The picturesque board or rail fence comes to mind first for many homesteaders. It won't work for goats, unless it's all but solid, because they can slip through openings you wouldn't believe.

Woven wire is less expensive, but that has drawbacks too. If your goats have horns, they'll put their heads through the fence, then be unable to get free. Even worse, they'll stand on the wire, or lean against it, until it drops to the ground and they can nonchalantly walk over it. Even with close spacing of posts and proper stretching, woven wire will soon sag from the weight of goats standing against it and will look unsightly, and eventually be useless.

Avoid barbed wire whenever possible. It's awfully ugly stuff around tender-skinned, big uddered goats.

Avoid picket-style fences where a goat can stand against the fence with her front feet, slip, and impale her neck on or between pickets.

The ideal goat fence would probably be chainlink, but like most ideals it just isn't within reach of most of us. A good and somewhat less expensive substitute is "stock fencing", which is made of welded steel rods and comes in a variety of lengths and heights. For goats it should be four feet high, and sections 16 feet long are easy to handle because the stuff is fairly light-weight. Stock fencing can be attached to regular steel or wooden fence posts. There are other forms of fencing that fall somewhere between the heavy duty stock fencing and the flimsy (for goats) field fencing, that will provide acceptable service.

Electric fencing should be used much more than it is for goats. They have to be trained to respect it, but once trained it's possible to fence large areas at low cost. A single strand about 30 inches from the ground will do the job once they've learned to keep their distance. Train them in a small area. Until they get zapped once or twice they'll be crawling under, jumping over, and just plain busting right through.

A good fence for fairly large areas consists of regular field fencing (woven wire) with a strand of electrified wire running just inside it. The electric fence keeps the animals from standing on and breaking through the field fencing, but the field fencing offers more security than the hot wire alone.

Good fencing is a necessity for goats, and fencing any sizeable area will be a large investment. It would be wise to plan on not pasturing your animals until you get some experience with their eating habits and their fence-destroying habits, and then go ahead based on your own experience.

6

Feeding

No aspect of goat raising is more important than feeding. You can start out with the very finest stock, housed in the most modern and sanitary building, but without proper feeding your animals will be worthless.

The proper feeding of goats requires special emphasis for several reasons. Paramount among them is the fact that many people who start to raise goats have little or no experience with any farm animals. Feeding goats is far different from feeding cats and dogs. Goats are ruminants, which affects their dietary needs, and unlike the average cat or dog they are productive animals, which puts additional strain on their bodies and requires additional nutriment.

A discussion of feeds can be very long, or very short.

It can be short if you buy commercially prepared feed and follow the directions on the label. It will have to be long if you mix your own because that will require at least an awareness of the basics of nutrition, physiology, bacteriology, math, and more.

There is no middle ground. You can't, for example, arm yourself with a "goat recipe" and do a good job of feeding. For one thing, the feed value of hay and grain varies from place to place and year to year, being affected by soil fertility, climate, and other factors. Pacific coast-grown grains are lower in protein than those grown elsewhere; hay harvested at the proper stage and well cured will vary dramatically in nutrients from hay that is cut too late and bleached or spoiled by improper curing.

More importantly, any given ration depends on locally-available ingredients and their comparative prices, and the suggested rations almost invariably have to be adjusted. Unless the feeder knows what to look for, the carefully formulated suggestions will be thrown out of balance by indiscriminate substitutions.

Likewise, even the homesteader who feeds commercial feeds can destroy the balance of the ration by haphazardly adding "treats" or by making use of available grains in addition to the commercial ration. You can no more prepare a balanced diet by adding a scoop of this to a handful of that than you could expect to use the same method to make a cake.

Fortunately, many goat raisers have an advantage over the average American when they start to learn about feeding livestock. They raise goats because of their interest in nutrition for their own bodies. They want raw milk, or milk from animals fed organically. So they already know a great deal about nutrition and its importance, which means they only have to apply some of that knowledge to goats.

Since feed prices started soaring in 1972, interest in home-mixed feeds has increased just as dramatically. Much of this interest no doubt stems from the illusion that the high cost of feed is due to the mixing. A quick look at the local prices of grain and supplements should convince most that buying the ingredients separately will effect little or no savings.

There is also a great deal of interest in "organic" feeds . . . feeds grown without chemical fertilizers, herbicides, or pesticides, and processed without medications, antibiotics, preservatives and artificial flavors. (Yes, artificial flavors are added to livestock feed, too!)

Then there are people who want to grow all or part of their goats' rations themselves.

To feed a dairy goat intelligently it will be helpful to know something about the animal's digestive system. People familiar only with human diets and perhaps those of dogs and cats should especially examine the process of rumination, for goats are ruminants. Like cows and sheep, they have four stomachs.

The process of rumination serves a very definite purpose, and has a bearing on the dietary needs of the animals. Ruminants feed only on plant matter which consists largely of cellulose and other carbohydrates and water, which makes adaptations in the structure and functioning of the stomach and intestines necessary. We commonly speak of "four stomachs" but in reality the large rumen (or paunch); the reticulum, and the omasum (manyplies) are all believed to be derived from the esophagus, while the fourth stomach, the abomasum or true stomach, corresponds to the single stomach of other mammals.

Vast numbers of protozoans and bacteria live in the rumen and reticulum. When food enters these "stomachs" the microbes begin to digest and ferment it, breaking down not only protein, starch and fats, but cellulose as well. The larger, coarser material is periodically regurgitated as the cud, rechewed, and swallowed again. Eventually the products of microbial action (and some of the microbes themselves) move into the "true" stomach where final digestion and absorption take place.

No mammal, including the goat, has cellulose-digesting enzymes of its own. Goats rely on the tiny animals in their digestive tracts to break down the celloluse in their herbiverous diet. You might say you're feeding the microbes and the microbes feed the goat, for without them grass and hay would have no food value.

Let's back up a bit to take another look at these stomachs, for not only are they of obvious importance to the goat: the goat owner (or at least the kid raiser) has some control over their development.

Watch a newborn or very young kid sucking. She stretches her neck out to get her milk. Due to the stretching process the milk goes past a slit in the esophagus, by-passing the first two stomachs and ending in the omasum. Here it is mixed with digesting fluids and passed to the fourth stomach, or abomasum.

Contrast this with a pan-fed kid, especially one fed only two or three times a day instead of four or five, and who is therefore more hungry and greedy. It must, first of all, bend down to drink rather than stretch upwards. Some of the milk slops through the slit in the food tube and falls into the first stomach, the rumen, where it doesn't belong. There is nothing else in this compartment since milk is the only feed consumed. There is no bulk. Gas forms, and scours are likely to result.

The good goatkeeper will strive to keep milk out of the rumen by proper feeding. Moreover, he will work to develop the rumen and reticulum the way they should be developed by encouraging the kid to eat roughage at an early age. Here's why this is important.

A young milk-fed kid has about 30% of its stomach space occupied by the rumen and reticulum. At maturity, a well-developed doe has a rumen that occupies 80% of the stomach space and a reticulum that takes up 5%. The omasum is 8% and the abomasum is 7%. (As an illustration of why feeding requirements of ruminants differ from single stomach animals, note that a horse's stomach holds 12-19 quarts, while the four of a cow hold over 250 quarts!)

The rumen does not increase to this size without proper stretching or development. Early feeding of roughage is essential.

Now let's examine a mature doe. She takes little time to chew. Notice how she draws in her neck to swallow, allowing the food to slip through the slit in the esophagus to the rumen. A slight fermentation begins as the microbes go to work. When at leisure, the goat brings up some of this material by regurgitation and "chews her cud." This time the mastication process is thorough. Now she *extends* her neck and the cud goes to the third stomach, or omasum.

Set a pail of water in front of the goat. Notice how she extends her neck to the far side to drink. This insures that the fluid goes to the omasum where it belongs, not to the rumen.

The goat must have a well-developed rumen to function properly, and a bulky diet to keep the rumen working right. This means hay (or other roughage) forms the basis for the diet.

This can be carried a step farther. In his book *Goat Husbandry* (Faber and Faber, London, 1957), David Mackenzie maintains that bulk is necessary for good milk production. He points out that milk production in British goats dropped 20 percent in 15 years . . . and by 12 percent just in the four years following the "derationing" of animal feedstuffs in 1949.

The reason, he believes, is that when concentrates were rationed during the war years the official concentrate ration for a milking goat was adequate if she had plenty of bulk food such as hay and roots. The allowance for kids and young stock was much more restrictive, and milk for kids very much so. His charts show a steady increase of milk production based on about 3,000 records from the British Goat Society, and a dramatic *decrease* after rationing was lifted. His conclusion was that excessive feeding of milk and concentrates to young goats prevents full development of the rumen.

The rumen, therefore, is of prime importance, along with the tiny animals that inhabit it. Next problem: feeding those tiny animals.

They are conditioned to what they've been accustomed to "eating." Change their diet and they can't cope. The result is a sick goat. Therefore, make any feed changes gradually. Many a goatkeeper has fed his goat an armload of cornstalks salvaged from the garden after harvesting sweet corn, and when the goat gets sick or dies he blames the cornstalks. In reality the problem was one of overload. Feed such delicious things sparingly, along with the regular diet, and everybody—protozoans, bacteria, goat and you—will be happier.

In addition to the special needs of the goat relative to rumination, it's important to feed her as a *dairy* animal. Production of milk requires more protein than would be needed for body maintenance, for example. So a milking doe is fed a ration of at least 16% protein, while a dry mature doe or buck will do well on 12%. Protein is expensive, and any excess is just wasted. You want to make sure the diet has enough, but not too much. Dairy animals also have a greater need for calcium and certain trace minerals.

It would be very helpful to think in terms of minimum daily requirements so familiar to the human diet when feeding goats. If you do, you'll be less tempted to stake the animal in a brush patch and assume she's "fed". She has no more nourishment in that situation than you would have if you lived on candy bars and soda pop.

Yet, there are people who treat goats like that . . . or toss them an armload of grass clippings and an ear of corn . . . or who mix up a concoction of corn and oats. The only difference is one of degree. (Lawn clippings, incidentally, are rich in vitamin A. Some people cure them like hay, and goats love them.)

Actually, there are no "minimum daily requirements" listed for goats. Very little research has been done with goats, certainly in comparison with the more economically important livestock like cows and hogs. But we can make certain assumptions based on cows and sheep.

All feeds contain water, organic matter, and mineral matter or ash.

Water is vital to life, of course, but it's also important in feed formulations because the quantity of water in various plants affects their place in the ration. Dry grain, for example, may contain 8%-10% water. Green, growing plants may contain 70-80% water. An animal fed the succulent plants ingests an enormous amount of water in order to get nutrients. Under certain conditions, bloat results.

Of the plants' dry matter, about 75% is carbohydrates, the chief source of heat and energy. These carbohydrates include sugars, starch, cellulose and other compounds.

The sugars and starch are easily digested and have high feed value. Cellulose, lignin, and certain other carbohydrates are digested only with difficulty and therefore it takes energy to digest them: their feed value is correspondingly lower. (This is one reason goatkeepers prefer "fine-stemmed, leafy green hay". The fine stems means less lignin and hard-to-digest material.)

If you buy feed, the carbohydrates are divided into two classes on the feed tag: crude fiber (or just plain fiber), and nitrogen-free extract. Nitrogen-free extract is the more soluble part of the carbohydrates, and includes starch, sugars, and the more soluble portions of the pentosans and other complex carbohydrates. It also includes lactic acid (found in milk) and acetic acid (in silage). Oddly enough, nitrogen-free extract also includes the lignin, which has a decidedly lower feeding value than cellulose.

Feed tags also list "fat", which actually includes fats and oils. They're the same except that fats are solid at ordinary temperatures while oils are liquid. In grains and seeds, fat is true fat. In hays and grasses, much fat consists of other substances. Many of these are vital for life, including cholesterol; ergosterol (which can form vitamin D) and carotene, which animals can convert into vitamin A.

The proteins and other nitrogenous compounds are of outstanding importance in stock feeding: many discussions on goat feeding focus on protein to the exclusion of everything else.

Proteins are exceedingly complex, each molecule containing thousands of atoms. There are many kinds of protein, some more valuable than others. (Livestock feeders speak of the "quality" of protein.) All are made up of amino acids, and protein must be broken down into amino acids before it can be absorbed and utilized by the body. There are at least 24 amino acids, but since they can combine like letters of the alphabet, there could be as many proteins as there are words in the dictionary.

The protein in plants is concentrated in rapidly growing parts—the leaves—and the reproductive parts—the fruits, or seeds. In animals, protein comprises most of the protoplasm in living cells and the cell walls. So it's important for muscles, internal organs, skin, wool or hair, feathers or horns, and it's an important part of the skeleton.

Protein, or crude protein, includes all the nitrogenous compounds in feeds. Protein is obviously essential for life, and is of extreme importance to the animal caretaker. Protein requirements vary among classes of livestock, being higher for young growing animals, reproduction, and lactation. And because protein is the most expensive portion of livestock feed, it should be added carefully.

"Ash" indicates the mineral matter of the ingredients. Minerals in plants come from the soil, but the mineral content of animals is higher than that of plants. Calcium and phosphorus are particularly important since they are the chief minerals in the bone and in the body. The body contains about twice as much calcium as phosphorus.

Other minerals are needed in trace amounts . . . but they are vital. Iodine, for example, prevents goiter; iron is important for hemoglobin which carries oxygen to the blood; copper, which is a violent poison, is also a vital necessity in trace amounts, as a lack of iron, copper or cobalt can result in nutritional anemia.

Other trace minerals are potassium, magnesium, manganese, zinc and sulfur.

Net energy values of livestock feeds are expressed in therms instead of calories. Since a therm is the amount of heat required to raise the temperature of one thousand kilograms of water one degree centigrade, one therm is equal to 1,000,000 calories.

Nutrients are constantly being oxidized in tissues to provide heat and energy. This oxidation maintains body heat and powers all muscular movements. Since the digestion of roughages requires more energy, it follows that one pound of Total Digestible Nutrients in roughages will be worth less than one pound of TDN in concentrates, which will not use up so much of its energy just being digested. TDN (total digestible nutrients) refers to all of the digestible organic nutrients: protein, fiber, nitrogen-free extract, and fat. (Fat is multiplied times 2.25 because its energy value for animals is approximately 2.25 times that of protein or carbohydrates.)

"Digestible" of course refers to nutirents that can be assimilated and used by the body. For this reason protein or crude protein is different from digestible protein. Digestible nutrients are determined in the laboratory by carefully measuring the amount of feed consumed and analyzing its content, and then analyzing the waste products. (Animal feces are largely undigested food, in contrast to human feces, which have a larger proportion of spent cells and other true "waste".)

As far as can be determined, no feed has been used for digestion tests on goats. For all practical purposes, results of tests on sheep and cows are considered accurate enough.

One more important consideration in feeding: vitamins. Vitamins were largely unknown before 1911, and there is still more to learn. But as of now, the only two of any consequence to goats are A and D.

Vitamin A is of prime importance to dairy goats because it's necessary for growth, reproduction and milk. It is of less importance in maintenance rations. Vitamin A is synthesized by goats that receive carotene in their diets. The chief sources are yellow corn and leafy green hay. Common symptoms of vitamin A deficiency are poor growth, scours, head cold and nasal discharge, respiratory diseases including pneumonia, and blindness. A severe lack of vitamin A prevents reproduction or weak (or dead) young at birth.

The other important vitamin for goats is vitamin D. As with other animals, lack of this vitamin causes rickets, weak skeleton, impaired joints and poor teeth. Vitamin D is necessary to enable the body to make proper use of calcium and phosphorus. The best and chief source of vitamin D is sunshine, but it is also available in sun-cured hay.

The B-complex vitamins are manufactured in the rumen and therefore the feeder has no concern with them directly. Vitamin E seems to have no special application to goats. Vitamin C is synthesized. (Only humans, monkeys and guinea pigs lack the ability to manufacture vitamin C.) Vitamin K is also synthesized.

For a more complete technical discussion of feeding, read *Feeds and Feeding* by Frank B. Morrison. First published in 1917, this book has been the authority in its field since that time. Dairy goats are not specifically treated in the book, but the information on cows and sheep is helpful.

With this very brief background we can begin to formulate a goat ration.

Composition of Goat Feeds

The main tool used is a list of the protein content of the feeds available. (See pp. 54–55.) The idea is to combine the various ingredients at your disposal in such a way that the combination will contain the desired amount of protein, or more accurately, digestible protein. But of course protein is only one aspect of feed value, so we must also bear in mind the other feed requirements, including palatability and cost as well as minerals, vitamins, fiber, and so on.

Because goats are ruminants, the main portion of their diet is roughage. So let's begin with that.

Roughage can be green, growing plants; alfalfa or clover hay; carbonaceous hays (timothy, brome, Johnson grass, etc.); corn stover (dry corn stalks); silage; comfrey, sunflower or Jerusalem artichoke stems and leaves; tree leaves, bark and twigs;

Average Composition of Selected Goat Feeds

	Crude Protein	Digestible Protein	Fat
Alfalfa hay	15.3%	10.9	1.9
Bermuda grass	7.1	3.6	1.8
Birdsfoot trefoil	14.2	9.8	2.1
Brome	10.4	5.3	2.1
Red clover	12.0	7.2	2.5
Mixed grass	7.0	3.5	2.5
Johnson grass	6.5	2.9	2.0
Soybean, early bloom	16.7	12.0	3.3
Timothy, early bloom	7.6	4.2	2.3
Succulents			
Green alfalfa, early bloom	4.6	3.6	0.7
Bermuda grass pasture	2.8	2.0	0.5
Cabbage	1.4	1.1	0.2
Carrot roots	1.2	0.9	0.2
Kale	2.4	1.9	0.5
Kohlrabi	2.0	1.5	0.1
Mangel beets	1.3	0.9	0.1
Parsnips	1.7	1.2	0.4
Potatoes	2.2	1.3	0.1
Pumpkins (with seeds)	1.0	1.3	1.0
Rutabagas	1.3	1.0	0.2
Sunflowers (entire plant)	1.4	0.8	0.7
Tomatoes (fruit)	0.9	0.6	0.4
Turnips	1.3	0.9	0.2
Barley	12.7	10.0	1.9
Steamed bone meal	7.5	—	1.2
Buckwheat	10.3	7.4	2.3
#2 dent corn	8.7	6.7	3.9
Linseed meal	35.1	30.5	4.5
Cane molasses	3.0	—	—
Oats	12.0	9.4	4.6
Field peas	23.4	20.1	1.2
Pumpkin seed	17.6	14.8	20.6
Rye	12.6	10.0	1.7
Soybeans	37.9	33.7	18.0
Sunflower seed, w. hulls	16.8	13.9	25.9
Wheat (average)	13.2	11.1	1.9
Wheat bran	16.4	13.3	4.5

Fiber	Nitrogen Free Extract	Mineral Matter	Calcium	Phosphorus
28.6	36.7	8.0	1.47	0.24
25.9	48.7	7.0	0.37	0.19
27.0	41.9	6.0	1.60	0.20
28.2	39.9	8.2	0.42	0.19
27.1	40.3	6.4	1.28	0.20
30.9	43.1	6.5	0.48	0.21
30.5	43.7	7.5	0.87	0.26
20.6	37.8	9.6	1.29	0.34
30.1	44.3	4.7	0.41	0.21
5.8	9.3	2.1	0.53	0.07
6.4	12.2	3.1	0.14	0.05
0.9	4.4	0.7	0.05	0.03
1.1	8.2	1.2	0.05	0.04
1.6	5.5	1.8	0.19	0.06
1.3	4.3	1.3	0.08	0.07
0.8	6.0	1.0	0.02	0.02
1.3	11.9	1.3	0.06	0.08
0.4	17.4	1.1	0.01	0.05
1.6	5.2	0.9	—	0.04
1.4	7.2	1.0	0.05	0.03
5.2	7.9	1.7	0.29	0.04
0.6	3.3	0.5	0.01	0.03
1.1	5.8	0.9	0.06	0.02
5.4	66.6	2.8	0.06	0.40
1.5	3.2	82.1	30.14	14.53
10.7	62.8	1.9	0.09	0.31
2.0	69.2	1.2	0.02	0.27
9.0	36.7	5.7	0.41	0.85
—	61.7	8.6	0.66	0.08
11.0	58.6	4.0	0.09	0.33
6.1	57.0	3.0	0.17	0.50
10.8	4.1	1.9	—	—
2.4	70.9	1.9	0.10	0.33
5.0	24.5	4.6	0.25	0.59
29.0	18.8	3.1	0.17	0.52
2.6	69.9	1.9	0.04	0.39
10.0	53.1	6.1	0.13	1.29

and root crops such as mangel beets, Jerusalem artichokes and carrots or turnips.

Green forages are rich in most vitamins except D and B_{12}. But if the animal is grazing it is getting sunshine and vitamin D, and ruminants can synthesize B_{12}. Rapidly-growing grass is also rich in protein.

However, because of the high water content of succulent green feed (and roots, too), these are low in minerals. The lack of minerals combined with the high water content (an animal could *drown* before it got enough nutrients from really lush grass) means that such forage does not constitute an adequate diet by itself. And it can cause bloat.

Alfalfa or clover hay is considered the ideal for goats because of the high protein content and because these are rich in calcium, the most important mineral. Good alfalfa or clover hay is cut before full bloom, when nutrition is highest. It is suncured quickly. Rain, or slow curing in damp weather leaches nutrients out of hay. Good hay is fine-stemmed, bright green and leafy.

Most of the nutrition is in the leaves. Hay that is baled when it's too dry suffers much shattering and loss of leaves.

The importance of good hay can be illustrated by the fact that good alfalfa can have as much as 40 Mg. of carotene per pound while alfalfa that is bleached and otherwise of poor quality can have as little as four Mg. per pound. Poor hay may be difficult to distinguish from some straw, which is the plant residue left after the grain has matured and has been harvested. Straw contains much fiber and especially lignin, and is used as stock feed only in dire emergencies. It has no place in a goat feeding program. Straw is bedding.

The carbonaceous hays have less protein and less calcium than the legumes, and the deficiency must be made up in the concentrate or grain ration.

Other hay plants can include barley, birdsfoot trefoil, Bermuda grass, lespedeza, marsh or prairie grasses, oat or wheat grasses, soybeans, or combination of these. For the benefit of inexperienced farmers it will be well to point out again that hay is made by cutting green, growing plants, and drying or "curing" them in the sun. Wheat, barley or oats, for example, can be cut when young for hay. If allowed to mature, the nutriment goes into the grain and the stems and stalks become yellow and have little food value, and the plant that could have been hay becomes straw.

Good alfalfa has about 13% protein; timothy and brome are closer to 5%.

While roughages are the most important part of the diet of a ruminant, they alone do not provide all of the needed vitamins and minerals, nor do they provide sufficient energy. Alfalfa hay has about 40 therms (energy) per 100 pounds; corn and barley have twice that. Especially if carbonaceous hays are fed (5% protein) additional protein and calcium is required. Hays do not contain sufficient phosphorus. These missing elements are provided in the concentrate ration.

A mature goat will require anywhere from three to ten pounds of hay per day depending on type, quality, waste, and other factors.

The concentrate ration is often called the grain ration, but this can be misleading. Here's why.

For lactating animals the protein content of the concentrate ration should be about 16% if the roughage is a good legume. With less protein in hay, more must be added to the concentrate. For dry does, 12% protein is sufficient.

Corn has about 9% protein, and only 6.7% digestible protein. Oats has about 13% protein and 9.4% digestible protein. Therefore a mixture of equal parts of corn and oats would contain 11% protein or about 8% digestible protein. Clearly, these grains alone will not meet the demands of the growing or milking animal. Protein supplements in the form of soybean oil meal (sometimes listed as SOM), linseed or cottonseed oil meal, must be added.

Milking animals also require more salt than is needed for animals on maintenance rations. It is usually added at the rate of one pound per hundred pounds of feed.

Because of the need for bulk in the diet of a ruminant, a concentrate ration should not weigh more than one pound per quart. Bran is most commonly used for bulk. (Beet pulp is sometimes used for does, but if used for a long period may cause urinary calculi in bucks.)

And finally, since goats generally shun dusty ground feed

as is normally fed to cows, the grains should be fed whole, or crimped or cracked. Cows do not digest whole grains well. Whole corn goes in one end and out the other. Goats seem to have better powers of digestion.

As a general rule, the grain ration for goats should not weigh more than one pound per quart. The weight of grains varies with the quality, but this chart shows the average weight per quart of selected feeds.

Weights of Some Common Goat Feeds

Barley, whole	1.5 pounds per quart
Buckwheat, whole	1.4
Corn, whole	1.7
Linseed meal	0.9
Molasses	3.0
Oats	1.0
Soybeans	1.8
Sunflower seeds	1.5
Wheat, whole	1.9
Wheat bran	0.5

This is fine for the homesteader mixing his own grain because it eliminates the bother and expense of grinding. But it also means the fine ingredients (salt, bran, oil meal, and minerals if used) cannot be mixed into the grain. To overcome this, most goat feeds contain cane molasses. In addition to binding the ingredients and making the feed less dusty, molasses is a source of iron and other important minerals, it increases the palatability of the feed, and does fed ample molasses during gestation are less likely to encounter ketosis. Molasses contains about 3% protein, but none of it is digestible. In addition, there is evidence (at least in dairy cows) that excess molasses interferes with the digestibility of other feeds. The digestive processes attack the more easily assimilated sugars in molasses to the detriment of other feedstuffs. Even so, molasses is an important feed for goats.

At last we're ready to formulate a ration. Here are several base formulas to work from.

1: Ration for a milking doe fed good alfalfa hay.
   12.6% digestible protein.

Corn	31 pounds
Oats	25
Wheat bran	11
Linseed oil meal	22
Cane molasses	10
Salt	1

2: Ration for a milking doe fed good alfalfa hay.
   12.6% digestible protein.

Barley	40 pounds
Oats	28
Wheat bran	10
Soybean oil meal	11
Cane molasses	10
Salt	1

3: Ration for a milking doe fed non-legume hay.
   21.2% digestible protein.

Corn	11 pounds
Oats	10
Wheat bran	10
Corn gluten feed	30
Soybean oil meal	24
Cane molasses	10
Salt	1

4: Ration for a milking doe fed non-legume hay.
   21.2% digestible protein.

Barley	25 pounds
Oats	20
Wheat bran	10
Soybean oil meal	25
Linseed oil meal	15
Salt	1

5: Ration for dry does and bucks.
    9.8% digestible protein.

Corn	58 pounds
Oats	25
Wheat bran	11
Soybean oil meal	5
Salt	1

6: Ration for dry does and bucks.
    10.1% digestible protein.

Barley or wheat	51½ pounds
Oats	35
Wheat bran	12½
Salt	1

These rations, followed more or less faithfully, could be expected to produce good results. There will be minor variations, because the feed value of grains depends in part on variety, weather and climate, and the fertility of the soil which produced them. Most grains grown in the Pacific northwest are decidedly lower in protein than the same grains grown elsewhere. Old-fashioned open-pollinated corn has more protein than the hybrids in common use today, and so on.

However, there are more serious considerations. One is that certain ingredients may not be available in your locale, or others may be more common and therefore less expensive than those listed. Grains can be substituted for one another by using the chart showing protein contents.

You can determine the weight of protein in a given feed ingredient, and by working with batches of 100 pounds, merely move the decimal point two places to the left to get the percent of protein in any ration.

Let's use a small homestead farm which produces its own grain as an example. In 1974 the corn crop was almost a total failure due to the wet spring, summer drought and early frost. But other grains were available. Here's what the milking does were fed:

Feed	Weight	% Crude protein	% Digest- ible Protein	Pounds of Protein
Soybeans	20 pounds	37.9	33.7	6.74
Barley	15	12.7	10.0	1.50
Oats	20	13	9.4	1.88
Buckwheat	5	10	7.4	.37
Wheat bran	5	16.4	13.3	.66
Corn	10	9	6.7	.67
Linseed meal	10	34	30.6	3.06
Molasses	10	3	–0–	–0–
Salt	1	0	–0–	–0–

Total Digestible Protein: 14.88 pounds

Divide the pounds of protein into the total weight of the ration (14.88 ÷ 100) and the percent of protein is 14.9.

It should be noted that some rations you'll find elsewhere work with crude protein rather than digestible protein. Since there are no digestibility trials on goats, both methods have flaws. It may be easier to obtain figures on crude protein for locally-grown feeds from your extension office, in which case *all* the ingredients should be calculated on the basis of crude protein.

This leads us to another—perhaps the most important— reason why every goat owner should have at least a basic knowledge of feed formulations. Homesteaders, and goatkeepers in particular, are notorious for dishing out treats or making use of "waste". These are both admirable, but look what happens.

Assume a goat is receiving one pound of a commercial 16% (crude) mixture. Maybe it costs the owner $10 a cwt, and he can get corn for half that, or he grew a little corn for the chickens and has some extra. Or the goat just seems to "like" corn! So he decides to give the goat one half pound of the regular ration and one half pound corn. Here's what happens:

Goat feed	50 pounds	8 pounds protein
Corn	50 pounds	4.5 pounds protein
	Total	12.5 pounds protein

That 16% mixture drops to 12.5% protein. That might be enough for the goat to maintain her own body, but not to produce kids and milk.

The same thing happens when the animal is given garden "waste" or trimmings. Such fodder replaces roughage, not grain, but even then it can cause unbalancing of the diet because elements of hay, for instance, will be missing from most of the garden produce.

This is not to say that rations can't be manipulated or that the homesteader shouldn't make use of what's available or cheap. It must be done with a certain amount of knowledge and discretion, however.

With this principle firmly in mind, let's examine some of the common feeds homesteaders have available and show an interest in.

Soybeans deserve special mention because many people look at the price of the oil meals and wonder why the beans can't be fed whole. They can, with certain restrictions.

Soybeans contain what is called an "antitrypsin factor." Trypsin is an enzyme in the pancreatic juice which helps produce more thorough decomposition of protein substances. The anti-trypsin factor doesn't let the trypsin do its job, which means the extra protein in the soybeans is lost, not digested. The anti-trypsin factor can be destroyed by cooking and it isn't present in soybean oil meal.

However, rumen organisms apparently inactivate the anti-trypsin factor when raw soybeans are fed in small amounts. Current recommendations for dairy cattle are that the ration not contain more than 20% raw soybeans.

DO NOT FEED RAW SOYBEANS IF YOUR FEED CONTAINS UREA! The result will probably be a dead goat. Urea is not recommended for goats in any case, but many dairy feeds for cows contain it. So does LPS . . . Liquid Protein Supplement . . . which some feed dealers will try to sell you when you ask for molasses. Urea is a non-protein substance that can be converted to protein by ruminants, and some people do feed it to goats because it's less expensive than the oil meal protein supplements, but other goatkeepers have reported breeding problems with animals fed urea.

Most goats are raised on small farms or homesteads where grain and hay are not produced. Such places can still grow a great deal of goat feed if the basic principles of feeding are followed. You can "grow milk in your garden" by planting sunflowers (the seeds are high in protein and the goats will eat the entire plants); mangel beets; Jerusalem artichokes; pumpkins; comfrey; carrots; kale; turnips; and others. In addition, such "waste" as cull carrots and apples and sweet corn husks and stalks can be utilized in the goat yard. These are treated like pasture or silage: they replace part of the grain ration, but not all of it. Feed at least one pound of concentrates per head per day to milking animals.

Many people with more time than money and a keen interest in nutrition are also avid collectors of weeds for their animals. For example, dandelion greens are extremely rich in vitamin A, and nettles are high in vitamins A and C. Goats relish these and other common weeds. It's just about impossible to imagine a real farmer on his knees gathering dandelion greens for his livestock . . . but many a goat farmer can, and reaps healthier animals and lower feed bills.

This brings up a point of particular interest to those who want to mix their own feeds because they aren't satisfied with commercially-prepared rations.

No one plant has everything any animal needs for nutrition. Goats seem to enjoy variety more than most domestic animals. Many goatkeepers prefer to provide food from as many different plant sources as possible to enhance the possibility that their animals are getting the nutrition they need, naturally, without synthetic additives. They like grain mixtures of at least five or six different ingredients.

This isn't as "efficient" as modern agricultural methods. Farmers know that alfalfa is rich in protein and calcium, both important to dairy animals. A great deal of feed can be harvested from one acre of alfalfa, and alfalfa hay has become the norm. There are even herbicides to kill weeds in alfalfa to keep it pure.

But almost any weed in your garden has more cobalt than alfalfa. And cobalt is required by ruminants to provide the bacteria in the digestive tract with the raw material from which to

synthesize vitamin B_{12}. Some, if not all, internal parasites rob
their hosts of this vitamin.

Alfalfa (and clover) has little cobalt because lime in the soil
depresses the uptake of this mineral, and lime is necessary for
the growth (and the calcium content) of alfalfa. Agribusiness
has found it more efficient to strive for high yields of alfalfa and
then add the trace minerals to the concentrate ration. Homesteaders who don't mind gathering "weeds" can meet their animals' nutritional needs naturally . . . and without the cash outlay
required for commercial additives.

Organic farmers have known this for years, of course, but
when their beliefs were confirmed by scientists in 1974 the idea
was hailed as "revolutionary". Researchers at the University of
Minnesota compared nutritive value and palatability of four
grassy weeds and eight broadleaf weeds with alfalfa and oats as
a feed for sheep, which have roughly the same requirements as
goats.

Lamb's-quarters, ragweed, redroot pigweed, velvetleaf and
barnyard grass all were as digestible as alfalfa and more so than
oats forage. All five weeds had more crude protein than oats and
four had as much as alfalfa. Eight were as palatable as oat forage.

One caution about weeds, though: don't gather them from
along roadsides where the lead content may be high due to
auto exhausts, and certainly not where spraying is done.

Also, under certain conditions, some weeds such as lamb's-
quarters and pigweed (and even "normal" crops such as oat and
wheat grass and sudan) can be toxic. When they are very young
or when they grow rapidly after a setback such as drought, they
can be dangerous. Some plants are also hazardous after being
killed by frost. Your county agent can give you more specific
information on these plants and conditions which are prevalent
in your area.

Tree trimmings fall into the category of weeds. Tree leaves
and bark are rich sources of minerals brought from deep within
the earth by tree roots. Although goats love pine boughs, there
have been reports that pine needles have caused abortions, so
caution is advised. They are rich in vitamin C, although goats
have no particular need for this vitamin, being able to manufacture it themselves. Also avoid wild cherry, since when the leaves

are wilted they are poisonous. Some weeds are poisonous too, of course, including milkweed, locoweed and bracken. Since these vary so widely in distribution, consult your county agent to see what's considered dangerous for cows in your area.

One more plant deserves special attention, because so many people are interested in it and because it's controversial. That's comfrey, or boneset.

There have been a rash of statements from county agents and state departments of agriculture knocking comfrey. Some of their reasons for *not* growing it are practical . . . for large farmers, not homesteaders. And much of their information is just plain wrong. All of their information came from a single release from the USDA which, upon checking, is unsubstantiated.

There are many goat and rabbit raisers who swear by comfrey as a feed, a tonic, and as medication for certain conditions such as scours. They can't all be wrong!

Even aside from that, comfrey should be in every goat owner's garden. It is high in protein, ranking with alfalfa, although there is some question about the digestability of the protein. But it is easier to grow and harvest than alfalfa, using hand methods. It is an attractive plant that even can be used for borders or other decorative applications: grow goat food in your front yard or flower bed! It has tremendous yields since it begins growing before alfalfa in spring and grows back quickly after cutting. And it is a perennial. It can be dried for hay although that entails a lot of work because of the thick stems. It must be cured in small amounts on racks rather than left lying on the ground.

No discussion of goat feeds would be complete without mentioning minerals. Most goat raisers supply mineralized salt blocks free choice, and also add dairy minerals to the feed.

While there is no sound research into the matter, there is some indication that this is unnecessary, expensive, and perhaps even dangerous. Too much of a good thing can be as bad as too little. Goats that are well-fed on plants and plant products from a variety of sources, grown on organically fertile soil, probably have little or no need for additional minerals.

There are certain exceptions. Plants grown in the goiter belt (from the Great Lakes westward) are low in iodine, and iodized salt will be good insurance. Certain areas of Florida, Maine, New Hampshire, Michigan, New York and Wisconsin and Western Canada have soils deficient in cobalt. Parts of Florida are deficient in calcium.

Phosphorus is a vital ingredient of the chief protein in the nuclei of all body cells. It is also part of other proteins, such as casein of milk. Calcium and phosphorus together comprise about 75% of the total mineral matter in the body, and about 90% of the bone structure, as well as 50% of the mineral matter of milk. Therefore it is of extreme importance to growing animals that are producing bone and muscle; pregnant animals that must digest the nutrient needs for the growing fetus; and for lactating animals which excrete great quantities of these minerals in their milk. Vitamin D is required to assimilate calcium and phosphorus. Also, the ratio of calcium to phosphorus is critical.

Roughages, especially legumes, are high in calcium, and grains are high in phosphorus. If these crops are grown on soils rich in these minerals the well-fed goat is likely to get enough of them.

A goat lacking phosphorus will show a lack of appetite, it will fail to grow or will drop in milk production if in lactation, and it may acquire a depraved appetite such as eating dirt or gnawing on bones or wood. In extreme cases stiffness of joints and fragile bones may result.

However, overfeeding calcium can be dangerous, too, especially for young animals. Lameness and bone problems can result later from excess calcium.

Iron is 0.01 to 0.03 percent of the body, and is vital for the role it plays in hemoglobin which carries oxygen in the blood.

Copper requirements are about one-tenth those of iron, and in greater amounts copper is a deadly poison.

Nutritional anemia can result from lack of iron, copper or cobalt. (This is different from pernicious anemia in man.) But it's very rare.

Other trace minerals are potassium, magnesium, zinc and sulfur.

If you feed your goats well, a trace mineral salt block will last a long, long time. In that case there is no need for adding minerals to the feed.

It can be seen that feeding is a science . . . and an art. Goats are not "hayburners" or mere machines to be fueled haphazardly. You wouldn't burn kerosene in a high-powered sports car, and you can't get the full potential from a goat fed improperly.

Summing up: feed your goats one pound of concentrate for maintenance and one pound extra for each two pounds of milk produced, along with all the hay they will eat. Some will do better on less, others will want more: that's the art, or part of it. "The eye of the master fattens the livestock."

Remember these basic concepts:

• The ration should come from as many different sources as possible, and fertile soil.

• Avoid sudden changes in feed which result in overloading the rumen bacteria and microbes.

• Pay attention to protein levels as well as vitamin and mineral content of the plants and grains you feed.

• Treat each animal as an individual, for they have different needs according to age, condition, production, and personal quirks.

7

Grooming

Goats require a minimum of care . . . but that doesn't mean they require *no* care. The goat keeper will quickly learn to disbud, tattoo, clip, and trim hooves.

Let's begin with the most important: hoof trimming.

The horny outside layer of the goat's hoof grows much like your fingernails, and must be cut off periodically. Gross neglect of this duty can cripple the goat. But it's really a simple job, and for the small herd won't take more than a few minutes a month.

How often you trim depends on several factors. Sometimes hooves grow faster than other times, or there are differences among animals. Goats living on soft spongy bedding will need more hoof attention than goats allowed to clamber on rocks or other hard ground. (One goat book claims that if a good-sized rock is placed in the goat pen, the animals will stand on it and keep their hooves worn down. All the people I know of who tried it said it didn't work . . . but it still sounds good, so maybe it'll work for you.)

There are several methods of hoof trimming, requiring different tools. The simplest is a good sharp jackknife. Others prefer a linoleum or roofing knife. Still others swear by the pruning shears (same ones you use for your roses) and there also are special shears made just for trimming hooves which works even better. For light trimming or finishing, many people like a Surform (a small plane with blades much like a vegetable grater).

These are some of the tools commonly used in hoof trimming.

Hoof trimming is easier if you have a helper, or if your milking stand is of a type that you can lock the goat in and still have room to get at all four feet.

With the goat secured, stand against her rear (tail to tail, as it were) and grasping one hind leg lift it up between your legs. Some goats don't seem to mind such acrobatics: others will protest rather violently. Keep a firm grip on it, and be exceedingly careful so she doesn't kick and injure you with the knife. For this reason many, ladies especially, prefer the shears rather than the knife, and it's a good idea to wear heavy gloves with either.

With the point of the tool, clean out all the manure and dirt imbedded in the hoof. If the hoof has not been trimmed in some time, it will have grown underneath the foot and can contain quite a lot of crud.

Next trim off those folded-over edges and obviously excess sidewalls. (You'll notice that dry hooves can be very hard. They're easier to trim after the goat has been walking in wet grass, and the shears cuts hard hooves better than a knife.)

Excess heel and toe should be trimmed off.

Then (here's where the Surform comes in handy) carefully

take thin slices from the entire bottom of the foot. The portion near the toe invariably needs more of this work than the heel. You can cut quite safely until the white portions within the hoof walls look pinkish.

Let her stand on the foot and see how it looks. A goat with good hooves stands squarely. A kid a few weeks old has the ideal hoof you're aiming for.

Do the other hind hoof the same way. Then you squat down just behind the front leg and bring it up over your knee and repeat the process.

In extremely bad cases where the goat looks to be wearing pointed elf's shoes it may take several trimmings to get them back in shape. It's better not to cut too much at once . . . and if they're really bad, the goat isn't likely to stand around patiently while you finish. Take off as much as you can and finish up later.

Bucks have hooves too! These poor fellows are more likely to be neglected than the girls are, but they shouldn't be.

The other major grooming duty for goatkeepers is disbud-

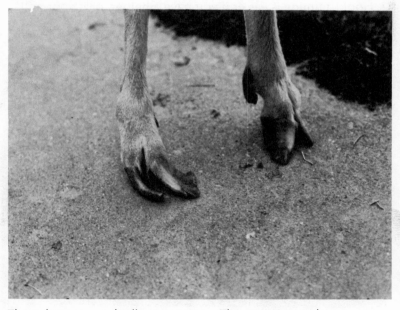

These hooves are badly overgrown. The goat is nearly crippled.

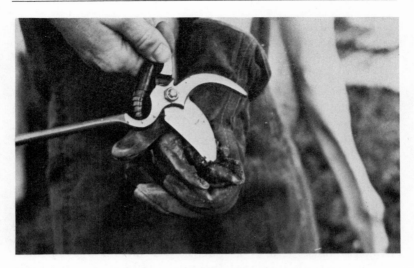

The first step is to clean out the accumulated manure and crud with a point of the nippers or a knife.

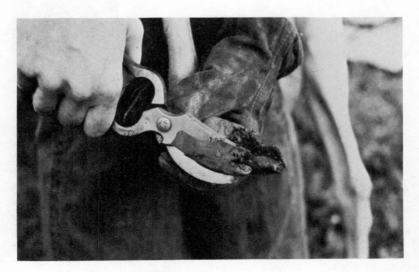

Then the excess sidewall is trimmed away.

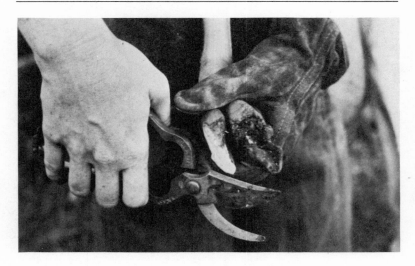

Excess skin over the heel and the point of the toe should
be trimmed away.

The hoof is shaved flat and even with a knife or Surform.

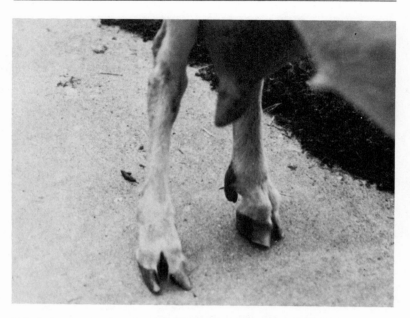

This is better . . . but it would be better if the doe hadn't
been allowed to go so long without a trim!

ding. Much has been written about the advantages and disad-
vantages of horns, and even more has been said around goat
barns. In summary, here is the gist of the arguments on both
sides:

For horns:

Horns are protection against dogs and other predators;
horns are beautiful and natural and a goat doesn't look like a
goat without them; disbudding is ghastly.

Against horns: they are dangerous to other goats and people,
especially children; it's impossible to build a decent manger to
accomodate a nice set of horns; horns are a disqualification on
show animals; horns really aren't much protection against dogs
—look to better facilities instead.

Disbudding involves destroying the horn bud on a very
young animal before the horns really start to grow. That's in
contrast to dehorning, which is the surgical removal of grown
or growing horns. Dehorning can be quite painful and even

dangerous to the goat, and so upsetting to the surgeon that even many trained vets won't do it, or at least not more than once. They certainly don't solicit the business.

Disbudding is relatively quick, easy and painless, although it may not appear so to the neophyte.

The recommended tool is the hot iron or disbudding iron. A kid-size disbudding iron can be made from a large soldering iron (with a point about the size of a nickel). The point must be ground flat. A calf disbudding iron works well. It can be used even after the horns have started growing, and in many cases used ones are quite inexpensive at farm sales. And of course they're available new, while none are made specifically for kids.

Hold the kid on your lap after the iron is hot enough to "brand" a piece of wood with little pressure. If the horn has not yet erupted or you aren't too sure of yourself, trim the hair around the horn button with a small scissors. Then, holding the kid firmly by the muzzle, press the iron into the button and hold it there to a count of 15.

There will be acrid smoke from burning hair, violent struggling (which isn't too violent with a kid weighing 8-12 pounds) and maybe some screaming. But when the 15 seconds are up everything will be back to normal except maybe your heartbeat.

Console the kid and compose yourself while the iron heats up again, then do the other horn button. Offer her a bottle of warm milk after it's all over and she'll forget all about it. And remind yourself that the next one will be easier.

I know one hardy, elderly homestead-type lady who didn't

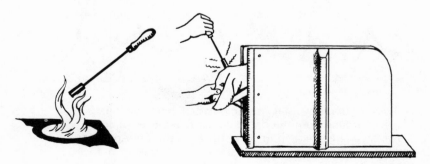

Left: Disbudding iron. *Right:* Kid-holding stall.

have electricity and who heated a metal rod in her wood burning stove for disbudding.

Another method, less hair-raising but also less successful
and potentially more dangerous, is to burn the horn buds with
dehorning paste, which is a caustic. There are several types and
brands available from farm supply stores and mail order houses.

With this method, the hair is clipped around the button and
Vaseline is applied *around,* not over, the area. Then the caustic
is applied. Kids treated in this manner should be isolated for half
an hour . . . one lady holds the kid on her lap while she watches
television . . . so no other kids will lick at it and so it doesn't rub
the caustic and get it on other parts of its body. The stuff can
cause blindness if it gets into eyes, and it will be quite painful on
other parts of the body.

Disbudding horns by using caustic potash. *From upper left:*
Clip hair around horn button. Then cut two pieces of adhesive tape large enough to cover the horn button. Apply
vaseline around the adhesive. Finally, remove adhesive and
apply caustic.

Bucks have much more stubborn horn buds than does, and there is also a difference in breeds. If scurs start to develop, merely heat up the iron and do the job over again. In some ways scurs (thin, misshapen horns) are more dangerous and troublesome than horns. They can curve around grotesquely and grow into an animal's head or eye, and thin ones will be broken off repeatedly resulting in pain and loss of blood. Kids' horns grow much sooner than calves': the directions on caustic were written for calves. Ignore them, as regards the time to do the job. The best time is when the kid is only a few days old.

Mature horns can be sawed off, usually with a special wire blade. They must be removed close to the skull, actually taking a thin slice of the skull with it, or the horns will grow back. There will be a great deal of blood, and obviously a mature animal is much more difficult to control. This is a job for a vet, and as mentioned, most of them don't even want to tackle it. (The recommended anaesthetic for goats is Rampon.)

There is an alternative, which apparently brings mixed results.

If very strong rubber bands are placed tightly around the base of the horn, the horn will atrophy and fall off. Some people file a notch in the horn very close to the skull to keep the band down where it belongs, and others claim putting tape over the band holds it on.

The problems arise when the horn structure begins to weaken. A goat may butt another, or merely get the horn caught in a manger or other obstacle, and break it off. If it's not really ready to come off there will be considerable pain and a great deal of bleeding. (Blood-stopping powder is a good thing to have in your barn medicine cabinet at all times anyway. If you don't have any, in an emergency a handful of cobwebs will do the job.)

Check the rubber bands regularly to make sure they haven't broken or moved.

With proper management . . . perhaps isolation of the animal so treated and the removal of obstructions and of course frequent inspection . . . this rubber band method seems clearly preferable over sawing, in most cases.

If you raise registered animals, you'll have to tattoo them.

If you don't raise registered animals, it's still a good idea to tattoo them. Tattoos are permanent identification numbers which can help in your record keeping, they can identify a goat long after you sell it, and in some cases they have helped retrieve lost or stolen animals.

You'll need a tattoo set, available from some farm stores, mail order houses and small-animal equipment dealers. Get a one-fourth or 5/16 inch die. Use green ink: it shows up even on dark colored animals.

You need a helper or a kid holding box, or fasten older animals in a stanchion or milking stand.

First, clean the area to be tattooed with a piece of cotton dipped in alcohol or carbon tetrachloride. LaManchas are tattooed in the tail web; other goats in the ears. Stay away from warts, freckles and veins. (I like this story: "Tattoo in the *ears!*" people ask Judy Kapture, *Countryside's* goat editor. "Doesn't it hurt?" She just waggles her dangling, pierced earings at them!)

Next smear a generous quantity of tattoo ink over the area. Paste ink can be applied from the tube; liquid ink requires a small brush such as a toothbrush.

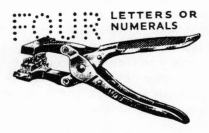

Tatoo marker

Then place the tattoo tongs in position and puncture the skin with a firm quick squeeze. (Be sure to test the numbers on a piece of paper first: they're backwards, like printer's type.) On very thin-earred kids the needles may go all the way through the ear. Gently pull the skin free. Some tattoo outfits have an ear release feature which eliminates this problem.

Next put on some more ink and rub it in thoroughly with the small brush.

It will take about a month to heal thoroughly. If you can't see the tattoo numbers, try holding a flashlight behind the ear in a darkened building.

You can tell the age of a goat by her tattoo. The ADGA letter for 1974 is H; in Canada it's F.

Very shaggy goats should have their hair trimmed, especially around the udder. It's a good idea to confine trimming to the udder region in the winter, in the interests of clean milk, but the entire animal can be clipped in the spring to keep them cleaner and cooler and to discourage parasites.

Electric clippers are nice, but new ones are expensive. Hand clippers (they're made for dogs) work well. The one-goat homestead could probably use a scissors on a really shaggy animal but it sounds dangerous to me.

One other goat-care skill is castrating. Buck kids saved for meat should be castrated by the time they're a few weeks old. A buck three months old is capable of breeding his sister, and no buck kept or sold as a pet should be left unaltered!

The most highly recommended method is the burdizzo, a tool that crushes the cords to the testicles and renders them ineffective. Simple, quick, and sure, but it requires a burdizzo. The small homestead isn't likely to have enough need for one to make the investment.

For castrating with a knife, have a helper hold the kid by the hind legs, his back to the helper's chest. Make a quick, clean incision with a *sharp* knife, remove the testicle and pull it out. Do the other one, and spray the wound with antiseptic.

It's also possible to castrate with rubber bands or emasculators, but this is easy to botch. Slip a strong very tight rubber band over the scrotum, making sure the testes are in the sack and not *above* the rubber band. The testicles will atrophy. It's not necessary to castrate buck kids for meat, but if you intend to keep them for more than three months and don't have separate facilities it's a good idea nevertheless.

8

Health

There seems to be a great deal of interest in goat ailments even though it's often said that goats are among the healthiest of domestic animals. Most people, who pay attention to proper feeding and other management details, have very few health problems with their goats.

The more goats you have and the longer you raise them, the more likely you are to see problems in your herd. The following information might be of some help then. But it goes without saying that an ounce of prevention is worth a pound of cure! Sickness is only an absence of health: health is the natural state, and can be maintained by proper nutrition and environment.

If your animals get sick, it's because of wrong conditions of feed, environment, or in some cases breeding. Treating the symptoms will help for the short run, sometimes, but unless the underlying causes are corrected any money spent on medication is wasted.

What's worse, many "illnesses" have purposes, and by "curing" them we're compounding the problem. Scours or diarrhea is one example. It's common in kids, and can result from feeding too much milk, cold milk, or from using dirty feeding utensils. You don't want to stop the diarrhea cold because that's nature's way of getting rid of the toxins, or poison. So you let it take its course while removing the *cause*: the excess milk, the cold milk or the unclean utensils.

We have been led to believe that germs are bad per se. Nothing could be farther from the truth. Even most pathogenic organisms will have little or no effect on a healthy body; only when the host is weakened because of some other factor—such as poor nutrition—does the pathogen get out of hand. Some bacteria are apparently harmless and some are actually necessary. It has been said that even if man had the technology to produce food and other necessities without animals of any kind, including soil organisms, he would still require the "animals" that live in and on his body—bacteria. TV commercials give just the opposite impression, that all germs are ugly and dangerous. Most people believe this and carry it to ridiculous and even dangerous extremes.

Your job, then, is to maintain the natural state of your goat's health by providing her with the proper feed and environment.

This is, unfortunately, easier said than done. Homestead goats are far removed from their natural surroundings and lifestyle and feeding habits. Even if we try to approximate these as closely as possible, as for instance by laboriously gathering mineral-rich weeds and tree branches, the goat is still eating under unnatural conditions and will take to the wrong foods as readily as the Eskimo has taken to sugar . . . to the detriment of both. Even milking is unnatural, because we demand much more of the doe than her kids would in the wild.

Against this backdrop, let's look at some of the most common health problems.

Abortion   Research indicates that goats are not susceptible to Bang's disease (Brucella abortus), even though Bang's disease tests are required almost everywhere goat milk is sold. Goats do abort, however.

If abortions occur in early pregnancy the cause is apt to be liver fluke or coccidia. Liver flukes are a problem only in isolated areas such as the Northwest where wet conditions favor them. Coccidia can be transferred by chickens and rabbits, both of which should be kept away from goat feed and mangers. (See internal parasites.)

Abortion is more common in late pregnancy. The cause can be mechanical, such as the pregnant doe being butted by an-

These goat fetuses were aborted.

other or running into an obstruction such as a manger or narrow doorway.

Certain types of medication can cause abortion, including worm medicines and hormones such as are contained in certain antibiotics. Medicate pregnant animals with caution.

Abscess From all accounts this is the most prevalent goat health problem. An abscess is a lump or boil, usually in the neck or shoulder region, which grows until it bursts and a thick pus is exuded. Dr. Samuel Gus, one of the few respected authorities on goat health in the U.S., has called a goat with a discharging abscess "a hazard to other goats and to humans". Any animal with an abscess should be isolated. The milk from such an animal should be pasteurized, Dr. Gus claims, and if the abscess is on the udder the milk should be discarded. Don't feed it to kids: dump it. The lump will become the size of a tennis ball or even larger, and burst by itself or it can be lanced when it's ripe. A small x-shaped cut will heal better than a straight cut, and the

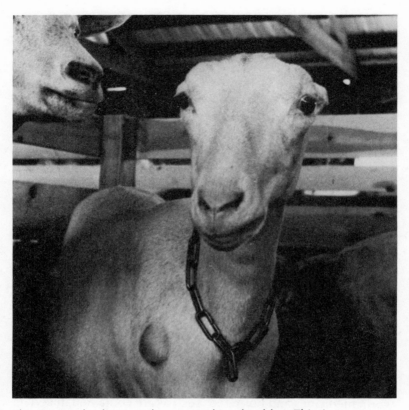

This LaMancha has an abscess on her shoulder. This is
probably the most common goat ailment.

incision should be made low on the abscess to promote drainage. Squeeze out the pus and burn the material and all cloths,
etc. that come in contact with it. Isolation and strict sanitation
are especially important during the period of drainage. If the abscess is caused by lymphadenitis it will have cheese-like pus; if
the pus is like mayonnaise it indicates pseudopneumonia. The
wound should be treated with iodine.

There have been recent reports of goat herds being vacci-
nated for corynebacteria and becoming abscess free in the process. However, this needs more study.

Herds that are free from abscesses generally stay that way
until a new animal is brought in or the goats come in contact

with others in some other way. We haven't had an abscess in our herd for over three years—except on one buck who is loaned out during the breeding season.

Bang's Disease (Brucellosis)   Brucellosis strikes fear into every dairyman, vet and public health official. Since it was discovered in 1887 in goats (and called "Malta Fever" after the place of its discovery) it is commonly associated with goats.

USDA statistics for 1971 reported 13 cases of Brucellosis in goats in three herds located in Arizona, Indiana and Ohio. A check with officials in Arizona and Indiana showed that their cases were in fact clerical errors. The one goat in Ohio was classed as positive on the test, suspicious on the retest, and after being slaughtered and subjected to a tissue test, negative.

Goats that have come up suspect after a Bang's test have invariably been pregnant or recently freshened. Subsequent tests are negative.

Of the 2,000,000 goats slaughtered under Federal Meat Inspection from 1960-65 and 1969-71, not one case was found.

Yet most new goat owners worry about TB and Brucellosis and most, if not all, veterinarians back them up. *Countryside's* veterinary columnist, Dr. C. E. Spaulding, puts it this way: although Bang's and TB are very rare in goats in the U.S., it is still safest to have them tested. "Your goat could be the only Bang's goat in the country but that wouldn't make the milk safe!" If it makes you sleep better have the goats tested. And to make absolutely sure, pasteurize the milk. Most of us will continue to take our chances, just as we do with cancer, fast-moving trucks, and falling trees.

Bloat   Bloat is an excessive accumulation of gas in the rumen and reticulum resulting in distension. If you've just turned the goat out on a lush spring pasture or if she found out how to unlock the door to the feed room, suspect bloat. As always, the best cure is prevention. Feed dry hay before letting animals fill up on high-moisture grasses and clovers. Don't feed great quantities of succulents such as green corn stalks if the animals aren't used to them.

Bloat is caused by gas trapped in numerous tiny bubbles,

making it impossible to belch. A cup of oil—corn, peanut or mineral—will usually relieve the condition. In extreme cases it may be necessary to relieve the gas by making an incision at the peak of the distended flank, midway between the last rib and the point of the hip and holding the wound open with a tube or a straw.

Colds   Some people claim vitamin C cures colds in goats and other livestock as well as in humans . . . and others say vitamin C doesn't do all that much for humans either. (Goats can synthesize vitamin C.) Decide for yourself.

Cuts   Cuts, punctures, gashes and other wounds can almost always be avoided by good management. They can be caused by such things as barbed wire, horned goats, junk or sloppy housekeeping and other conditions under the control of the goatkeeper. Clean such wounds with hydrogen peroxide and treat with disinfectant such as iodine. Use your own judgment to decide if stitching is required or get the animal to a vet.

Cystic ovary   This condition is indicated by irregular, infrequent or absent heat periods. Since the ovaries cannot function, the animal cannot be bred. Dispose of her.

Goat pox   The symptom is pimples that turn to watery blisters, then to sticky and encrusted scabs on the udder or other hairless areas such as lips. It varies in severity.

   Pox can be controlled by proper management, especially involving sanitation. Infected milkers should be isolated and milked last to avoid spreading the malady to others. Time and gentle milking are the best cures.

   Very similar conditions are caused by irritation. I have seen cases caused by dirty, urine-soaked bedding and by use of udder-washing solutions that were too strong. In these cases the cure is wrought by removing the cause. An antibiotic salve will keep the skin supple and prevent secondary infections. Traditional treatment is methyl violet to dry up the blisters, but this is very drying and can make the udder painful.

Enterotoxemia   The usual symptom of enterotoxemia is a dead goat. There is always misery, and almost always a peculiarly evil-smelling diarrhea. With some strains there may be bloat, or staggering.

Enterotoxemia is also called pulpy kidney disease ond over-eating disease. An autopsy soon after death will often show soft spots on the kidney.

It is caused by a bacterium which is always present, but which, when deprived of oxygen in the digestive system, pro-duces poisons. The proper conditions can be induced by over-feeding. Goats build up resistance to the poisons produced in small, regular amounts, but they can't handle sudden surges of them.

There are six types of Clostridium perfringens bacteria which cause enterotoxemia. Types B, C, and D cause the most trouble, with type D most often affecting sheep and goats. Anti-toxin can be administered if you get there fast enough, but death is usually swift. Where enterotoxemia is a problem, vaccines are available from your vet. Annual booster shots are required, and the kids will get antibodies from their dams. The best pre-vention is proper feeding on a suitably bulky, fibrous diet.

Johnne's Disease   This is one of the "wasting" diseases. Fre-quently the only symptom is skinniness or loss of condition. Scouring is a typical symptom in cattle but not always in goats. It cannot be diagnosed accurately in goats except by autopsy. The intradermal Johnin test used on cows has come up nega-tive on goats which actually had the disease as determined by autopsy.

The disease apparently infects the young, either by inter-uterine transmission, congenitally at birth or by mouth. If the disease follows the pattern it does in cows, adult animals can be sources of infection even if they do not show clinical disease. The kid infected at birth typically won't start getting sickly for $1^1/2$ to 2 years.

There is no reliable test, and no cure. Prevention consists of starting with a clean herd and keeping it that way.

Lice   Suspect lice if your goat is abnormally fidgety and has a dull scruffy coat. Fresh air and rain are good preventatives. Ask your vet for a louse powder approved for use on dairy animals. For an old-time cure, apply a mixture of two parts lard to one part kerosene.

Mastitis   Symptoms: a hot, hard, tender udder; milk may be stringy or bloody. Mastitis may be subclinical, acute or chronic. Hard udders (usually just after kidding) that test negative for mastitis are referred to as congested, and usually disappear. In mastitis the aveola or milk ducts are actually destroyed. Since it is necessary to identify the bacteria involved, the services of a vet are required.

Mastitis can be caused by injury to the udder, poor milking practices, or by transference by the milker from one animal to another. Teat dips have proved of great value in controlling the disease among cattle, although the solutions must be diluted for goats.

Congested udders are best cleared up by letting the kids suck and massage the teats and udder for 3-4 days after parturition.

The California Mastitis Test is available for the goatowner to conduct himself. It's available from vet supply houses. Some people have claimed success in treating mild cases by feeding vinegar or baking soda.

Milk Fever (Parturient Paresis)   Symptoms: anxiety, uncontrolled movements, staggering, collapse and death. Usually it occurs within 48 hours of kidding. It's caused by a drastic drop in blood calcium, which is related to the calcium level of feed consumed during the dry period, and even to incorrect feeding of young animals. It can be brought on by sudden changes in feed or short periods of fasting. Curing milk fever requires quick action and a vet, who will administer calcium borogluconate intravenously.

Poisoning   Symptoms are vomiting, frothing, and staggering or convulsions. Because of the nature of a goat's eating habits, poisoning from plants is rare: she takes a bit of this and a taste

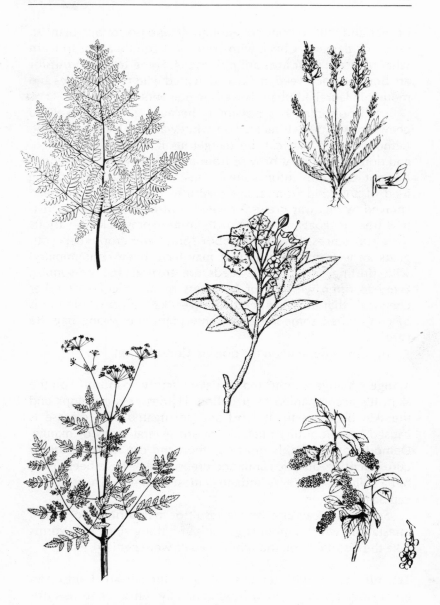

Five poisonous plants. Upper left: Brake fern. Upper right: Loco weed. Center: Mountain laurel. Lower left: European Hemlock. Lower right: Wild cherry.

of that and will seldom eat enough of one poisonous plant to do much damage. Check with your local county agent to learn what plants in your area are poisonous. Some to watch out for are bracken, locoweed, milkweed, wilted wild cherry leaves and mountain laurel. Rhubarb leaves are poisonous.

Lead poisoning is a possibility because goats are likely to chew on wooden surfaces. Use whitewash or lead-free paint, (although even that can be dangerous if enough is ingested) and don't gather goat browse from roadsides where concentrations of lead from automobile exhausts are heavy. Avoid taking feed of any kind from along roadsides that might have been sprayed by highway crews for weed control. Don't feed Christmas trees to goats, as many of them are sprayed with various toxic substances. If your neighbor sprays any crop, keep your goats away from any area that may have been contaminated with drifting spray. Many seeds are treated and poisonous. Every so often we hear of fertilizers or insecticides or other chemicals that look like feed additives, killing off whole herds of cows when someone mistakenly grabs the wrong bag. Be careful.

Antidotes depend on the poison. Consult a vet.

Mange   Mange is indicated by flakey, scurfy "dandruff" on the skin. It's accompanied by irritation. Hairlessness develops and the skin becomes thick, hard and corrugated. The disease is caused by a very tiny mite. There are several types of mange. Demodatic is probably most common and can be stubborn to cure. Nicotine, arsenic sulfur and creosote dips have been used, but a solution of 0.06% lindane is most common. These agents can be dabbed on.

Scurfy skin can also be the result of malnutrition (or worm infestation, which amounts to the same thing since the worms take the nutrition you thought the goats were getting).

Tetanus   Goats with tetanus will have their heads held up in an anxious posture and will be generally tense. There is difficulty in swallowing liquids, and muscular spasm. Death occurs within nine days. Tetanus or lock-jaw requires a wound for the germ to enter, but it can be something so simple the goat keeper doesn't even notice it. Treat all punctures and cuts with iodine, and pay special attention to the navels on new-born kids. If you have horses, or if horses or mules have been on your farm in the past 20 years, tetanus vaccination is recommended. Curing tetanus is a job for a vet.

Ketosis   (Ketosis includes pregnancy disease, acetonemia, twin lambing disease and others). This ailment occurs in the last month of pregnancy. The doe becomes dull and listless and goes off feed. In advanced stages she will become paralyzed and may abort. Death results. To test for ketosis get some "Ketostix" or "Lastix" from your vet or local drugstore. These are chemically treated and are used to test the urine. If she has ketosis the sticks change color.

Ketosis affects overly fat animals . . . but also underfed ones in poor condition. It can be brought on by extreme exercise.

Treatment consists of administering a cup of molasses twice a day, or one-quarter pound of brown sugar twice a day, or six ounces of glycerine or propylene glycol daily. Even with treatment mortality is 50%.

Worms   Because of their proportion of intestines to body size, goats are particularly vulnerable to worms. A worm-free goat would be impossible to find except in a laboratory, and chances are she'd die from the condition. A wormless goat is not whole.

David Mackenzie makes some very interesting observations on worms in his book, *Dairy Goat Husbandry*. By superimposing a graph showing seasonal liability of parasitic worm infestation on a graph of the seasonal variation in the metabolic rate of adult goats, he theorizes that worm infestation plays a helpful role in carrying out seasonal changes in metabolism and digestion. "The wormless goat," he concludes, "is an ideal to be cherished by the salesman of veterinary medicines, and by no one else."

Little or nothing has been done in the way of research along this line of thought, and since "worms" have as nasty a connotation as "bacteria", most dedicated goat raisers worm at least twice a year (spring and fall) and many worm every three months.

Again, domestic goats are not raised under natural conditions, and in some cases worming *is* necessary. For example, no goat would frequent the sodden and close-cropped lands where liver flukes present a danger if it weren't forced there by man. Close grazing on any land allows even the natural parasites of the goat to get out of control. On the other hand goats fed in mangers are unlikely to come into contact with these parasites.

Goats wormed routinely, whether they need it or not, have no opportunity to build up resistance. If they should be subjected to serious infestation they'll come off a lot worse than goats that haven't been wormed as a matter of course.

An interesting corollary to this is the fact that those who worm regularly must switch back and forth between vermifuges: the worms build up a resistance to the poisons.

Tobacco leaves do not cause wormlessness, nor does garlic. But these can be used as deterrents if you subscribe to the idea that a manageable number of worms may not be harmful, and may actually be desireable.

Don't misunderstand: worms can kill a goat. Goats with severe infestations caused by poor management must be treated while you improve the management. But feeding a goat worm poison every time you trim her hoofs "just to be sure" is like giving yourself a shot of penicillin every morning just to make sure you don't encounter some malady during the day.

If you're the kind of person with a medicine chest full of remedies for your own body you'll read this chapter with a far different attitude than people who don't even use aspirin. In that case, the recommended wormers for goats are Tox-i-Ton and Thibendizol.

Goats are normally healthy animals. Some of them would be a lot healthier if people just left them alone. Maintain cleanliness and sanitation, provide proper housing, make a study of feeding, and not only will you not have to become a veterinary expert: you and your goats will have a lot more fun.

9

The Buck

Since goats won't produce milk without kidding it follows that they must be bred. That requires the services of a buck and that entails a whole 'nother look at goat keeping.

Beginners are usually advised to forget about keeping a buck. There are many reasons. No doubt the most practical one is in terms of expense. A buck requires the same amount of housing, feed, bedding and grooming as a doe. Therefore if you have one doe and one buck the cost of your milk is double what it would be without the buck. The point at which a buck becomes profitable depends pretty much on your own situation. If you live in such a remote place that there is absolutely no stud service available within a "reasonable" distance (and this might be a few miles or a few hundred miles, again depending on you) then it might well be most economical to have your own buck. Expensive, but less expensive than the alternative.

But even then all your problems aren't solved, at least if you're going to do right by your goats and your own future in the avocation.

In the first place, good bucks are expensive. In recent years $100 for a week-old buck kid has been a common price. *Good* bucks are worth it, of course, as the buck you use this year will affect every goat in your herd a few years down the line, whereas each doe will only produce one or two doe kids per year. The buck is truly "half the herd."

A good breeder won't sell you a "cheap" buck: if it isn't good enough to improve most goats of its breed it is slaughtered at birth or castrated. Unfortunately there are too many people raising goats who are not good breeders.

This no doubt deserves some explanation, because there are two approaches to raising livestock, and not only are most beginners somewhat puzzled by the differences, but their natural inclinations frequently attract them to the less desirable approach.

We're speaking here of breed improvement. While most emphasis on breed improvement naturally comes from people who are involved in showing their animals, be they rabbits or dogs, cows or goats, there is more than ample evidence to prove that the "commercial" or homestead producer has every bit as much to gain from striving for improvement . . . and perhaps more. And there is really very little . . . or nothing . . . to lose.

I have found it extremely frustrating to deal with people who place little or no emphasis on breed improvement, or who even

Erik Borg

A Toggenburg buck

A Nubian buck

actively belittle the fancy-pants show enthusiasts as if their interests were somehow contradictory. Nothing could be farther from the truth.

For proof we need only turn to the commercial dairy (cow) farmer. Almost invariably these practical, tough-minded, cost-conscious farmers use the best purebred, registered bulls they can find. They may not have the slightest intention of ever showing a cow or of raising purebred cattle (although more of them are finding that purebred cows are practical for the same reasons purebred bulls are practical). They use purebreds because it pays off in the milk pail.

Milk production per cow has just about doubled since the last century. While some of that is due to feeding practices, surely a large share of the credit must go to genetics.

No such progress has taken place in goats. It may not even be desirable. But no lover of goats can deny that there are entirely too many half-pint milkers around (and being sold to unsuspecting novices who have heard that goats give a gallon of milk a day). The reason is simply poor culling practice, often starting with the selection of the buck.

Now, it's true that showing and in fact the entire registration system, sometimes aids and abets this: there *are* goats that place high in the showring but that don't produce enough milk to pay for their grain ration. It IS true that too many goats are registered (and sold for high prices and allowed to reproduce more high-priced goats) simply because they are purebred. If you just want milk you can ignore the faults in the system. But you won't be producing milk efficiently if you ignore the system. Ignore the *faults,* not the entire system.

That means follow the system as it's meant to be followed. Breed the best to the best and cull the rest. Cull means destroy: it doesn't mean passing on substandard animals to the first sucker who comes along or selling them as pets only to have them find their way back into someone's dairy. Too many people take the short-term view of the economic loss incurred, and as a result shortchange themselves, and future goat keepers, for the long term.

Your chances of improving your herd are practically nil if you breed your does to the neighbor's nondescript pet buck simply because he happens to be cheap and available . . . or if you buy a buck just because he's cheap. You aren't going to milk the buck, but never forget that you're going to milk his daughters. If he doesn't have the genetic potential for milk production, his daughters won't have it either. If the buck is not much better than the doe, you aren't working for breed improvement. In fact, you're not even breeding goats: you're merely freshening does. In the cases where goatkeepers are only interested in freshening does, because they only want the milk and don't even keep replacement animals, the worthless buck would have little effect if the offspring were routinely butchered. In practice this seldom happens, and the downgraded goats are foisted upon a world that already has too low an opinion of these valuable animals.

So how do you choose a buck that will produce superior offspring? You start by examining his pedigree . . . the record of his ancestors. If it's milk you want, make darn sure there's milk in that pedigree.

While there admittedly are lovely grades that give milk by the ton, without pedigrees there is no way of knowing who

A Saanen buck

their ancestors were or how good they were. A pedigree and milk production records of several generations of forebearers may not be insurance, but they're a valuable management tool and much to be preferred over flying blind.

Now, let's suppose you purchase a fine buck of impeccable breeding, excellent health and ideal conformation. Are your problems over? Not quite.

If you have four does, a fair average for a homestead herd, you can expect four doe kids the first year. Chances are excellent that you'll want to keep one or more of them: after all, didn't you buy the high-powered buck to improve your herd? But then, it's evident that the next breeding season you'll be making father-daughter matings. This isn't bad. While many people attach the human-oriented stigma of incest to such matings, they are actually the surest way to breed improvement in the hands of an expert. But most homesteaders are not experts in animal genetics, if only because a herd of a few animals would require years and years to give the breeder the experience necessary to be an "expert". It's sufficient to say here that inbreeding emphasizes faults as well as good points; it's nothing to be dealt with haphazardly. So when your herd sire's first daughters come into heat you'll want another buck, which means that once again you'll have to decide whether you can afford to keep one buck for each two females. Actually, there is some evidence that inbreeding affects goats less than some other animals. But are your original goats good enough to be perpetuated—or should they be upgraded by outcrossing?

Somewhat akin to this it should be noted that no animal is "perfect." All have faults. It's the job of the breeder to eliminate those faults as much as possible in future generations, while at the same time preventing new ones from showing up. An illustration of this would be a doe with very good milk production but a pendulous udder. That udder fault is going to shorten her productive life, it will make her more liable to encounter udder injury and mastitis and so forth. So you'll want to breed such a doe to a buck that tends to throw daughters with extra-nice udders, in hopes that the offspring will have both good production and acceptable udders. They won't be extra-nice in view of the genes contributed by the dam: but they can be improved by proper buck selection.

The problem here is that, of four different homestead goats, there are likely to be four different faults! It's extremely unlikely that even an excellent buck will be strong enough in four different areas to compensate for all of them. From the standpoint of breed improvement then, each doe in your barn is very likely to be best matched by a different buck.

These are real and important and practical considerations. If they don't impress you we'll have to mention some of the other, more commonly voiced objections to keeping bucks.

Unlike does, buck goats *do* smell, especially during the breeding season. Girl goats and some goat people are inclined to like the aroma but not only your clothes but even your living room furniture will get to smelling like ripe billy goat, which is usually something less than desirable.

Of even more interest to people who are new to goats are

what they often call his "objectionable disgusting habits". Most city people are shocked when they find out that the cute and playful buck kid grows—very quickly—into a male beast who not only tongues urine streams from females (and makes funny faces afterwards) but who also sprays his own beard and forelegs with his own urine. This is natural goat behavior, but be that as it may, even many of today's broad-minded people find such behavior a bit much. Needless to say, the loveable buck kid loses a few friends when he reaches this stage.

Bucks are powerful animals—I've seen 2 x 6's snapped by them—and one that has not been raised properly or finds himself in an untenable position due to thoughtless handling or the taunting of children can be a dangerous animal. (I have never owned a buck that was any more hostile or aggressive toward humans than a doe—although they haven't been effeminate either, which would be a fault in a buck. But enough other people speak of "mean" or "dangerous" bucks that it seems likely they exist, and you should be forewarned.)

Because they are powerful, and because of their natural sexual instincts, a buck requires much more expensive and/or elaborate housing than the does, especially during the breeding season. They must be housed separately if only to avoid off flavored milk. And an inadequately-penned buck will soon be found with the girls.

In spite of all of this, there are many practical and logical reasons for keeping bucks even for small herds, and many people do. While there are many advantages to buying a proven sire—a buck you know is not sterile and who throws daughters with the traits you want in your herd—such bucks are either very expensive or old and otherwise worn out. Most people buy buck kids just off their dam's colostrum.

As mentioned, most good breeders dispose of buck kids at birth, even very good ones, because there is so little demand for them. Only the very best are kept, and almost invariably these have been reserved far in advance.

Buck kids are raised very much like doe kids. They grow faster, but take longer to fully mature. However, even though a buck may not stop growing until he's three years old, he is capable of breeding by the time he's three or four months old. Don't let his size fool you! This means separate penning is necessary almost from the beginning, and at least by two months of age.

A buck may be used for limited service even before he's a year old. Most authorities say he should be limited to 10-12 does his first year. A mature buck can service more than a hundred does per year according to some reports.

If you prefer to use outside bucks it's a simple matter to put the doe in the trunk of the car or the back seat even and drive off. Artificial insemination is still another possibility. Either of these has certain very definite advantages over keeping bucks on the homestead.

See the chapters on breeding and on grooming for more information.

10

Breeding

A goat obviously must be bred in order to produce kids and milk. Those 155 or so days between breeding and kidding are extremely important to the goat, and for the first-time goat-keeper especially, they are anxious ones.

It all starts with the doe's estrous cycle or heat period. Goats (and most other animals) "cycle"; that is, they are fertile only for relatively short periods at more or less regular intervals. Unlike cows or hogs, which come into heat year around, goats generally come into heat only in the fall and early winter. A doe will accept service from a buck only when she is in standing heat, usually. If she is not in heat, copulation won't result in pregnancy anyway because the sperm and the ova aren't in the right place at the right time.

Seasonal breeding has obvious advantages for animals like deer. Their young are born when it isn't too cold; there is plenty of lush, milk-producing feed for the mother and tender grasses and leaves for the young to be weaned on; and the offspring is fairly strong and independent by the time the weather gets harsh again. Wild goats were in this same position, but desirable as such an arrangement may be for wild animals it puts the goat farmer in a bind.

In an earlier chapter we examined a lactation curve. If you plot such curves for several goats, all of which have been bred more or less at the same time, it's apparent that the goatkeeper

will be drowning in milk during part of the year and dry as a
bone during another part. This is perhaps the most serious single
problem of commercial goat farming. If people want to buy goat
milk they want to buy it regularly . . . not just when it's plentiful.
In addition, the normal lactation curve is actually reinforced by
seasonal curves that are equally normal in both cows and goats,
due to feeding conditions, weather and other factors. Animals
simply give more milk in the summer than in winter. Add to that
the fact that more people *want* goat milk in the winter than in
summer, and it's easy to see that the poor goat farmer has a
problem.

To a lesser extent homesteaders share the same dilemma.
If you have just one goat, even if she has a lactation of ten
months, you'll be without any milk at all for two months of the
year. With two goats you can attempt to breed one in September and one in December. Then you will theoretically never be
without milk, but a look at the lactation curves plotted together
will show that your milk supply will be far from steady. You'll
have too little or too much far more often than you'll have just
enough. While this rightfully belongs in the chapter on what to
do with extra milk, the point is that you'll want to have your
does bred as far apart as possible but while still avoiding the
risk of having a doe miss being bred entirely. With some does,
in some years, even December may be pushing it: they simply
won't come into heat again until September. There *are* kids born
in every month of the year, but for the practical homesteaders
the out-of-season breedings are few enough to be insignificant.

For many beginners, and especially those with only one or
two goats, it's very difficult to tell when a doe is in heat. The
usual signs are increased violent tail wagging, nervous bleating,
a slightly swollen vulva sometimes accompanied by a discharge,
riding other goats or being ridden by them, and sometimes lack
of appetite and drop in milk production. If a buck is nearby
there will be no doubt: she'll moon around the buck pen side of
her yard like a love-sick teenager.

If you lack a buck and have trouble detecting heat periods,
or just want to make darn sure she's in heat before you make a
several hour trip, you might use this trick. Rub down an aromat-
ic buck with a cloth, or even tie one around his neck for a cou-

ple of hours. Gingerly stuff it into a canning jar and screw the lid on. When you suspect your doe is eager for male companionship give her a whiff of that cloth and your suspicions will quickly be confirmed or denied.

But you can't breed a doe with canned buck aroma. If you don't have a buck you'll have to load her in the car and take her to one. We've found the trunk to be the best place for a goat. She will not "disgrace herself", as one old puritanical goat book puts it, laying down. Furthermore, even if you aren't overly concerned about nanny berries in your vehicle, some goats tend to get carsick standing up and will be too woozy to be interested in breeding. Others love riding as much as dogs and delight in sticking their heads out the window . . . which could easily cause an accident on a busy freeway among drivers who have never seen a goat in a car.

There is another possibility which interests many people: artificial insemination. From the standpoint of breed improvement there is no better method of breeding, for several reasons. Anyone, anywhere, can use some of the best bucks available and at low cost. Bucks with standing fees of $100 are available through AI for as little as $10. Because of the length of time semen can be stored, does can be bred to bucks that have been long dead . . . and they often are, by the time their daughters prove themselves. Inbreeding problems mentioned in the chapter on bucks would easily be eliminated by A I. Even goatkeepers who aren't overly concerned about upgrading can readily see the advantages of not having to keep a buck or having to traipse all over the countryside with lovesick does in their cars.

But there are shortcomings to AI, too. Just a few years ago the main one was that does bred artificially didn't conceive . . . a decided liability. We've gained considerable experience since then, and there is every reason to believe that rapid progress in technique will continue to make AI more efficient and desirable. A conception rate of 75% was considered good as of 1973.

In most cases it will be necessary for the breeder to do his own inseminating. This will require a liquid nitrogen semen storage tank costing perhaps $300-$400, plus close to $10 a month to have it charged. That's pretty expensive if you only have a few goats, although considering that you could use semen from several different bucks and thus do the best job of upgrading, it really is more economical than owning your own sires. A more serious handicap is learning to do the inseminating.

It isn't difficult, and numerous seminars on AI for goats have been held around the country. But for most people the whole process seems just too technical and "medical" to be practical on a homestead basis.

There is no doubt that AI can do a great deal to improve the goats in this country, and my prediction is that it will become increasingly common. Since a description of methods and equipment is not only beyond the scope of this book—but also would be out of date much sooner than the publisher would appreciate—if the process interests you it would be a good idea to keep current through the goat magazines and associations.

A doe will be in standing heat for 24 hours, although this varies widely. If she is not bred, she will come into heat again in 21 days, although this too varies considerably.

If a doe is serviced and still comes back in heat, there may be several causes. She may not have been bred at the most opportune time. Generally one more try will do it. (It is not necessary to leave the buck and doe together for long periods: if the doe is really in standing heat, one service is sufficient, and that won't take more than a minute . . . which sometimes seems silly after you've just spent an hour on the road and still have to drive home again!)

Sterile bucks are rare, and if a buck is sterile obviously none of the does he breeds will conceive. However, sperm can have reduced viability at certain times due to over-use of the buck or other factors.

If a doe simply will not get bred the most common cause is cystic ovaries—a growth preventing ovulation—and she is worthless. Over-fat does are often difficult breeders because of a buildup of fat around the ovaries.

Another serious condition, although we don't know how common it is, is hermaphroditism, or bisexualism. A goat looks like a doe externally, but internally actually has male organs of generation. Not all "hermies" have obvious external abnormalities. Examine the vulvas of newborn doe kids carefully. A growth about the size of a pea at the bottom of the vagina is abnormal. Unusual behavior in a normal appearing doe kid is cause for suspicion. Hermies are often overly aggressive or unusually withdrawn.

The word hermaphrodite goes back to Greek mythology and the story of the son of Hermes and Aphrodite who became united with the body of the nymph, Salmacis, while bathing in her fountain. In goats, the condition is often related to the mating of two naturally hornless animals. Not only do the genetics get a little complicated, but many practical goat breeders claim that the geneticists are wrong anyway.

Basically you must decide whether a naturally hornless buck is homozygous or heterozygous, that is, whether or not it inherited a gene for horns from either of his parents. Theoretically, there can be no homozygous does because they'd be hermaphrodites and couldn't breed. Both types of bucks will produce some hermaphrodites when bred to hornless does, according to theory, but the homozygous hornless buck will produce more.

If either the buck's sire or dam were horned, he's heterozygous. If neither parent were horned, you can't be sure without seeing a number of his kids. If any of the kids have horns, the buck is heterozygous. If all of the kids are hornless, even out of horned does, chances are the buck is homozygous.

All of this is of intense interest to geneticists and large goat breeders and people who take a keen interest in breeding . . . but the average homesteader is much better off to follow the lead of the major commercial goat farms and just avoid hornless-to-hornless matings.

Doelings are sexually mature as early as three or four months of age. In most cases spring kids that are well-developed and healthy should be bred when they weigh about 80 pounds and are seven or eight months old, which means they'll kid at one year of age. Being bred too early will adversely affect their growth and milk production; being bred too late does not contribute to their health and welfare; it's expensive to keep dry yearlings; and records show the does which kid at one year of age produce more milk in a lifetime than those which are held over. Many people mistakenly hold back young does because "they look so small" or because seven months seems so young. With proper nutrition, they'll produce kids and keep on growing themselves.

The bred doe normally will still be milking, but advancing pregnancy will cause most goats to dry off. Some people who really want milk will continue milking a naturally drying off doe as long as she gives a few squirts: others figure even a pound isn't worth their trouble, just stop milking, and the doe dries off.

In any event it's good practice to dry off a doe two months before her kids are due. In most cases, simply quitting milking and reducing the grain ration will cause the animal to dry off naturally. In cases of extremely heavy milkers it may be necessary to milk her out at intervals. Milk once a day for a few days, then every other day, then stop. Reducing or eliminating grain will help dry off an animal.

No good dairy animal—goat or cow—can eat enough during lactation to support herself and her production. That's why she requires the rest to build up her body before peeling off her own reserves to fill your milk pail. It's been said that for each pound of increase in body condition during the dry period a Holstein cow will produce an extra 25 lbs. of milk, a Guernsey 20 lbs. and a Jersey 15. We could anticipate proportionate results with goats.

This is not to say a pregnant animal should be overconditioned or fat: just in good condition. In fact fat causes problems in pregnancy. But the goat should have a well-balanced diet. Too much feed produces kids that are too large to be easily delivered. Excess minerals in the doe's diet produces kids with too-solid bones, which also causes difficulty. A fibrous diet with rather low protein is ideal for the first three months of pregnancy when the kids are growing slowly.

Most of the kids' growth comes in the last eight weeks of pregnancy. During this period the ration should be changed gradually, not only because two-thirds of the kids' growth is taking place, but because this is when the doe needs to build up her own reserves for the next lactation. High protein still isn't required—about 12%—but there is definite need for minerals and vitamins, especially iodine, calcium, and vitamins A and D. Bulk such as provided by beet pulp or bran is required, and molasses will supply some iron as well as the sugar which will help prevent ketosis, and it will have a desirable slight laxative effect.

Finally everything is ready. The goat stork cometh.

11

Kidding

The "miracle of birth" is aptly named. Like all miracles it's invested with wonder, awe, excitement, and joy. There have been cases of people who would have nothing to do with goats —until they saw newborn kids frolicing in fresh clean straw and fell in love. (I'm married to one of them.)

There is little doubt that the first kidding season brings the new goatkeeper excitement that is hard to duplicate in today's plastic and artificial world. Most of them, judging from the mail I get and my own first experience, are scared silly as parturition approaches.

Most of this fear comes I believe from reading books and articles describing the things that can go wrong. You expect the worst. But goats have been having kids all by themselves for thousands of years. While there can be problems 99 per cent of goat births are completely normal and won't even require your assistance. The chances for a normal birth are enhanced by proper feeding and management during the latter stages of pregnancy.

The average gestation period for goats is 145 to 155 days. Some experts say there is evidence that goats and sheep can control the time of birth to coincide with proper weather conditions. In my experience they control it all right, but usually to have the kids and lambs arrive on the coldest, most miserable night of the year.

Start checking your animals frequently and carefully 140 days after breeding. When she is getting ready to kid the doe will become nervous and will appear hollow in the flank and on either side of the tail. There may be a discharge of mucous, but this may appear several days before kidding. When a more opaque, yellow, gelatinous discharge begins it's for real.

Kids can be felt on the right side of the doe. It's good practice to feel for them at least twice a day. As long as you can feel them there they won't be born for at least 12 hours.

Also if you feel of the doe regularly you'll be able to notice the tensing of the womb. After this, one of the kids is forced up into the neck of the womb, causing the bulge in the right side to subside somewhat. This will be noticeable only if you have paid close attention to the doe in the days and weeks before. The movement of the first kid also causes the slope of the rump into a more horizontal position. At this point you can expect the first kid within a couple of hours.

Many people look to "bagging up" or enlarging of the udder as a sign of approaching parturition. This is unreliable. Some goats don't bag up until after kidding, and others will have a heavy milk flow far in advance. In some cases, if the udder becomes hard and tight, it may be neccesary to milk out the animal even before kidding.

Although man strives to take good care of his animals, he often complicates things. We've mentioned feeding of the pregnant animal, which can affect the ease with which she delivers. A free-ranging, experienced goat knows what to eat, but if we must bring her her food she must depend on our judgement. Likewise the goat kidding outside on her native mountain range knows what to do when her time approaches. She is probably safer and in more hygenic surroundings on her mountain than in your barn. It's just about impossible to duplicate such conditions for domestic animals.

There are innumerable instances of goat owners going to the barn for morning chores and finding a couple of dried-off, vigorous and playful kids in the pen with their mother. But it's definitely preferable to have some idea that the kids are on the way and to make certain preparations for them.

The doe should have an individual stall for kidding. It should be as antiseptic as possible and well-bedded with fresh, clean litter. Something softer than long straw is preferred if you have it. Don't leave a water bucket in the kidding pen. It can be dangerous and the mother isn't thirsty at this point anyway.

As mentioned, 99 times out of a hundred there will be no problems and your assistance will be unnecessary and perhaps even unwanted. Normal labor can last anywhere from a matter of minutes to four hours or more.

In a normal birth the front feet and nose are presented first. It can be seen that this presents a more or less cone shape which gradually distends the vagina. In abnormal births the kid will try to be born with one foreleg bent back, effectively hooking it inside the womb; tail first; or in some multiple births the kids and umbilical cords get all mixed up in the womb.

The solution to these problems is to insert your disinfected, lubricated hand into the birth canal to find out what's wrong. (A germicidal soap will serve as disinfectant; mineral oil can serve as a lubricant.) If you've never seen a newborn kid this not only is scary: it's difficult to imagine what you're feeling for. Sort out the heads and legs and if necessary rearrange them for the proper presentation. In most cases it will be a simple matter to "lead" the first one out the next time the doe strains. Pull, but very gently, working with the doe: otherwise hemorrhage might result.

Chances are the others will come by themselves soon after.

A pessary, available from your vet, should be inserted in the vagina after manual exploration to minimize the danger of infection.

Does usually have two kids, but one, three, or even four and five aren't all that uncommon. If no more come within a half hour and the mother seems relaxed and comfortable, you can assume that's all there are.

If you happen to encounter a difficult birth—a dead kid that the doe can't expel, for instance—better get a vet or a sheep-raising neighbor to help. But experienced help is needed in less than one birth out of a hundred.

In most cases the umbilical cord breaks by itself. If it doesn't, tie it off with a soft cord (a shoelace is good) about two or three inches from the kid's body, and snip it off on the doe's side of the knot with a sharp scissors.

The umbilical must be disinfected. Iodine spray is convenient, but better protection can be had by pouring some iodine into a small container, pressing it up to the kid's navel, then briefly tipping the kid over to insure good coverage of the navel with the iodine.

Watch for the afterbirth. If the doe doesn't expel it within several hours get a vet: a retained afterbirth is nothing to mess around with. If it's just hanging out of the doe, don't pull on it. You might cause hemorrhaging.

When the afterbirth is expelled dispose of it. Some does eat it, a natural instinct most wild animals have developed to protect their young from predators that might otherwise be attracted. Most goatkeepers don't care for this sort of behavior, but it won't hurt the goat. Won't do her any special good though, either.

The newborn kids will be covered with mucous which the doe will remove by licking. You can help by wiping it off with a soft cloth, paying special attention to the nose and mouth (so the kid can breathe) and eyes.

Examine the kids for defects. If you do not intend to keep bucks for meat, it's much easier to drown them in a pail of water even before they're dried off. And remember, only the very best bucks from outstanding dams and sires should be kept for breeding.

Check for supernumary teats (extra teats). There are several variations of this condition, some of which make the animal worthless as a milker. If the extra teat is sufficiently separated from the main two it might not interfere with milking, and can even be removed at birth with a surgical scissors or by tying a fine thread very tightly around the base and letting it atrophy. Actual double teats make the animal worthless. Bucks may also have double teats, and such animals should not be used for breeding.

If the tip of the vulva has a pea-like growth the animal is a hermaphrodite and will not breed. It should be destroyed, and the mating that produced it should not be repeated. Some hermaphrodites are not obvious.

Unwanted kids, bucks and does can be butchered and dressed like rabbit. Wait until they dry off: they're easier to handle.

Put the kids in a clean, dry, draft-free place . . . a large box
works well . . . and turn your attention to the doe. She has lost
a tremendous amount of heat, even in warm weather. Offer her
a bucket of *hot* water. Most pet-type goats also get a special
treat: perhaps a small portion of warm bran mash or oatmeal or
a handful of raisins. Provide some of the best hay you have.

In very cold climates, a box like this will keep chill drafts
off newborn kids.

Perhaps the most common kidding problem encountered is entering the barn on a blustry spring morning to find a new-born kid cold and shivering. DON'T feel sorry for it and bring it in out of the cold. Place it in an enclosed box or pen padded

Goats can be trained to pull small carts—to the delight of youngsters or to the relief of homesteaders who want to move compost and other materials. Almost any book on dog training can be used to train goats.

with an old blanket or feed sacks, away from any hint of a draft, and with a heat lamp if the weather is really nasty. But don't let it get hot: the switch back to normal temperatures will be as dangerous as the cold that brought on the problem in the first place.

In the case of a severely chilled kid on the brink of death this might not be the solution. If you find one still wet and thoroughly chilled and almost lifeless, one of the best ways to save it is to immerse it up to the nose in a bucket of water at about 100 degrees. That's about the environment it just came from. When it has revived, dry it thoroughly, wrap it in a feed sack or blanket, put it in a box in a protected place and watch it carefully.

If you do end up taking a kid into the house for any reason during cold weather, you're stuck with a goat in your house until the weather warms up, and even then you should harden it off by degrees rather than exposing it to the cold all at once.

The kids probably won't be interested in eating for the first few hours, even though they'll be standing and walking. Their first food must be their dam's colostrum or "first milk" . . . a thick, sticky, yellow non-foaming milk. This contains important antibodies and vitamins. It's so important, in fact, that if you buy a goat just a short time before kidding and she has no opportunity to build up antibodies to your particular location and therefore can't pass that protection on to her kids through colostrum, you might possibly encounter problems with sickly kids. You can allow the kids to nurse or you can milk out the doe a few hours after the birth and feed the kids from a bottle or pan. Colostrum scorches more easily than milk. Warm it carefully.

If for some reason the doe has no colostrum you can make a fair imitation. Use three cups of milk, one beaten egg, one teaspoon of cod liver oil and one tablespoon of sugar.

Some goat raisers with fairly large herds freeze excess colostrum for emergencies. While generally considered not fit for human consumption, there is a cheese made from colostrum that some people consider a great delicacy. Most homesteaders won't have enough colostrum to worry about making use of it.

12

Raising Kids

The excitement of freshening over, your goat barn can settle into a routine. For the first three or four days your doe will produce colostrum, the thick yellow milk so necessary for the kids's well-being. After that there hopefully will be enough milk for both the kids and for you. And after two months of relative inactivity your goat barn will be a hectic place. In addition to the usual feeding and cleaning tasks, you'll be milking twice a day . . . and raising kids.

Raising kids right requires some knowledge and a lot of work. Different goat raisers have different opinions about how the job should be done, but none can deny that the first year of the goat's life . . . along with her breeding and prenatal care . . . is an important determinant in how she will behave and produce later.

If the doe has a congested udder or a very hard udder the condition often can be helped by letting the kids nurse for the first few days. The suggested procedure is to bring the kids to their dam every few hours, rather than leaving them together. While this entails more work it eliminates a lot of commotion and consternation later on. First fresheners, which often have very small teats, are also frequently left with their kids at first if the milker's hands are too large. The teats will enlarge with time.

If you milk the doe, do it within a few hours of kidding and offer the kids some colostrum. Feed milk at close to goat body temperature, which is 103 degrees. Colostrum scorches easily: bear this in mind when you heat it.

How will the kids be fed? Nursing is certainly the easiest method but not necessarily the best. Some people say it ruins the dam's udder, which is important not only if you intend to show her but also if you want milk production. Possibly a more important consideration is that you don't know how much the doe is producing or how much the kid is getting. Kids left with their mothers are much wilder than hand-raised kids. Another important consideration for homesteaders, is that once a kid learns to suck its dam it will be difficult, maybe impossible to teach it not to. Some does wean their kids relatively early, but there have been other cases where yearlings are still sucking . . . and your milk supply for the kitchen is lost. The only solution in those cases is complete separation. Much better to do it right away.

While some people advocate nursing, claiming it's "natural," they invariably have more milk than they can use or sell. The small homestead herd is rarely in such a position.

That leaves a choice between pan feeding and bottle feeding.

Here again, some breeders prefer bottle feeding as being more natural, pointing out that with pan feeding, where the animal is forced to lower its head, milk can get into the rumen because of the kid's eagerness and greed. The kids drink too fast. Milk does not belong in the rumen, and digestive upsets can result. For bottle feeding, lamb nipples are used on pop bottles or something similar.

Pan-fed kids are simply given milk in a pan or dish of some sort. In either case the utensils must be scrupulously clean, but pans are much easier to wash than bottles and nipples. Nubians' ears will get in pans and sometimes skin irritation will result.

A third alternative finding increasing use among breeders with large numbers of kids are devices such as the Lam-Bar and the Lamb Saver. These are larger containers with a number of nipples that allow you to feed a whole penful of kids at once and leave only one utensil to clean up.

Frequent small feedings are better than infrequent large feedings for any animal, and especially for young ones. For the first three days the kids should get four to eight ounces of colostrum four times a day, depending on size and appetite. Then, until they're eight weeks old, they get anywhere from eight to ten ounces three or four times a day and an equal amount of warm water afterwards, if they want it. Most will. As the kids learn to eat hay and grain the milk should be gradually decreased, and in most cases they can be weaned completely by reducing the amount and frequency of feeding by the time they're eight weeks old.

Some people feed milk for much longer—as long as six months in some cases. This certainly isn't necessary, and according to Dr. Leonard Krook of Cornell University it may actually be harmful. Kids overfed calcium (milk is high calcium) are likely to develop bone troubles in later life.

Actually there are no hard and fast rules about anything connected with kid raising. This is one area where the old saying, "The eye of the master fatteneth his cattle" is especially true. You don't want to starve the kids, obviously. But you don't want to kill them with "kindness" either. This happens most frequently by overfeeding milk and thus causing scours (diarrhea) which can be fatal. If a kid starts to scour reduce the milk. (Cold milk or unclean utensils can also cause scours.)

You don't want kids to become "fat and healthy" because fat *isn't* healthy. Strive for condition, not overcondition. The kid should be producing bone, not fat, to develop her full potential in later life.

Early development of the rumen is extremely important for later production, too. Most kids will start to nibble at fine hay by the time they're a week old. They should be encouraged to do so with kid-size mangers and frequent feedings of fresh, leafy hay.

The largest goat dairy in the United States, Laurelwood Acres in Ripon, Calif., with 1,200 animals, raises about 400 kids a year with some methods that break some of the accepted rules. They feed only twice a day, although they admit more frequent small feedings would be better where it's economically possible. The kids get a quart of milk at each feeding. Newborn kids don't drink that much, of course, but they do by the time they're two weeks old.

Laurelwood has tried milk replacers of several types and

didn't like any of them. One brand seemed to work well and the kids looked fine, but when the records were checked on these kids two years later most of them were dead or had been culled because they lacked stamina.

Doe kids are weaned at eight weeks, bucks at ten. This is when grain feeding starts. The grain is similar to a calf starter, which most goat raisers use. A half pound per kid is fed twice a day. At six months of age the kids are switched to a milking ration. By seven months the doelings weight 75-80 pounds and are bred.

Milk-fed kids weighing from 20 to 30 pounds are in great demand as meat in some localities, especially around Easter. In the early 1970's, such kids were worth $20 and more in New York.

Any bucks kept for meat should be castrated by the time they're eight weeks old. They're capable of reproduction by the time they're three months old. The best way to castrate is with a tool called a Burdizzo, which crushes the cords to the testes and renders the male inefectual. Some people also claim to have good success with elastrators: small, strong rubber bands placed near the base of the scrotum, which will then atrophy. To castrate with the knife, have a helper hold the kid by the hind legs. Make a quick clean incision with a very sharp knife and pull out the testicle. Repeat the process on the other one. With experience, it takes only a few seconds. Treat the incisions with antiseptic.

Bucks kept for meat require no special diet, but some chevon (goat meat) affecionados claim milk and browse produce the best meat. Butcher kids can be fed grain liberally.

13

Milking

The new goat raiser must learn about goat feeds and nutrition, about bucks and breeding and raising kids—all for one purpose only: to get milk. Milking therefore is at the apex of the pyramid of all goatkeeping skills.

Goats must be milked at regular 12 hour intervals and according to a regular routine for best results. Milking at 6 a.m. one day and 9 a.m. the next is one of the easiest ways to depress milk production. You might milk at 7 a.m. and 7 p.m. or at noon and midnight, but it should always be as close to 12 hours apart as possible and *always* at the same time.

Even more than regularity, sanitation is important. One of the main reasons for keeping goats is having milk better than any to be found in the supermarket dairy case. This requires not only a knowledge of dairy sanitation but a rigid adherence to sanitation principles.

Milking equipment can be simple or elaborate. You could, if you wanted, milk into a bowl, use an inexpensive small strainer and store the milk in the refrigerator in fruit jars. But if you intend to milk 730 times a year for the foreseeable future you will get satisfaction and better quality from the proper equipment.

A four-quart seamless stainless steel hooded goat milking pail is no luxury, despite its fairly high price (more than $20). I consider it a necessity. I know of one that's been in use for

This stainless steel milking pail (lower right) is specially
made for milking goats. It's completely seamless for easy
and thorough cleaning, and the half-moon cover and handle are removeable. (The cover helps keep dust, dirt and
hair out of the pail). The strainer, also made especially for
home dairies, is also seamless and stainless and uses disposable strainer pads to filter the milk before cooling.

The other utensils pictured are for holding milk. The
smaller cans (available in one, two and four quart sizes) are
adequate for homestead use and will fit in most refrigerators. The larger, 16-quart pails are more likely to be found
in small dairying operations. (Photo courtesy of Countryside General Store.)

more than 20 years. They're made especially for goats, so you'll have to check the goat supply houses or magazines to find one. Being seamless and stainless, the pail is easy to clean and disinfect. The hood and the handle are removeable to enable thorough cleaning. Some people milk in plastic buckets, but no amount of cleaning can get bacteria out of the pores in plastic and you'll soon end up with a product that's unfit for human consumption.

A strainer is a necessity. It must be of a type that uses disposable milk-strainer pads: I won't even comment on the practice of running milk through a hunk of cheesecloth. Small tin kitchen strainers work well with small amounts of milk, but a larger size that's ideal for goats is sold by the same people who sell the pail. Strainer disks or pads should be available at any farm supply store.

Milk can be stored in one-, two- or four-quart glass jars or in metal milk cans of the same sizes. Again avoid plastic if it's to be reused. Look for something easy to clean and sterilize.

In addition the goat dairy will require a bucket to hold udder wash; an udder sponge or cloths; drying towels or paper towels; and udder wash and utensil disinfectants and cleansers. A scale for weighing milk production is a good idea although obviously not necessary, and a strip cup will help you maintain a constant check on one aspect of your herd's health and the quality of the milk your family drinks.

Ideally the milking should be done in a special room, away from the goat pen, away from any source of dust or odor. It should have good ventilation, running water, electricity, a drain, a minimum of shelves or other flat surfaces that gather dust, and impervious floor, walls and ceiling so it can be kept cleaner than your kitchen. All this—and more—is required for commercial milk producers.

Ideals are often unattainable. Lots of people with just a few goats milk right in the barn (not in the pen) and I have yet to hear of any of them dropping dead from drinking the milk.

Actually, goat milk has less bacteria than cow milk as it leaves the udder. But on the debit side the goat milk is more likely to pick up coliform bacteria during the process of milking. This is due in part to the dry nature of goat dung. It actually becomes dusty. Combined with loose housing most commonly used for goats, this results in dung-dust and coliform bacteria in the air, on flat surfaces, and on the goat. The goat milker is more likely to disturb the hair on the belly than the cow milker just because of the size of the animal.

Whether milked in a parlor or in the goat barn, the goat should have long hair around the udder clipped, she should be brushed to remove loose hair and dust, the udder should be washed with a milk disinfectant and a hooded milking pail should be used.

We're finally ready for the actual milking. It looks easy . . . until you try it. But then with a little practice it IS easy, and you wonder why you had milk up your sleeves and all over the wall and your legs the first time you tried.

A milking stand is far more comfortable than squatting, especially if you have a number of milking does or if you tend to creak anyway. But it's not necessary. Either way you're at the goat's side facing toward the rear. Goats can be milked from

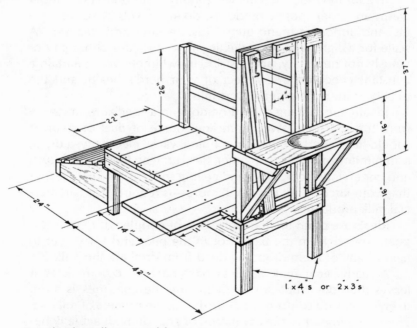

Details of a milking stand for goats

either side but they develop a definite preference for the side they're always milked from.

The first step is washing the udder with warm water and an udder-washing solution. Follow the directions for the proper strength. Strong solutions can cause udder irritation. Dry the udder and your hands with a paper towel—a fresh one for each goat—to avoid chapping and other udder problems.

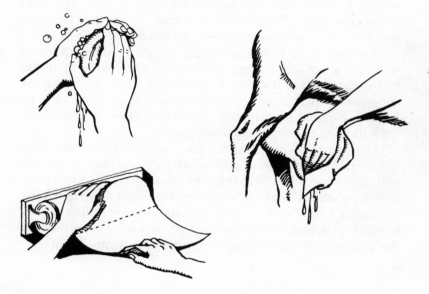

Upper left. In milking following the same routine at each milking. Be gentle. Strangers and the family dog tend to make a doe "hold up her milk." Keep them out of the barn or milk room. After milk pail and wash pail are ready, be sure your hands are clean. *Upper right.* Wash the goat's udder in chlorinated warm water—from 120°–130° F. The udder should be washed *just before milking.* Use a separate wash rag for each goat. *Lower left.* Have a roll of paper towels handy to milking stand. Dry the goat's udder and your hands. Wet hands can cause a chapped udder—and worse.

Then grasp the teat with your thumb and index finger encircling it near the base of the udder. Do not grasp the udder itself, which is sometimes tempting on goats without a clearly defined teat. That could cause udder tissue damage and the tissue can work its way into the teat with disastrous results.

Squeeze your thumb and index finger together to trap milk in the teat. This must be held firmly or when you squeeze the rest of the teat the milk will be forced back up into the udder rather than into your pail.

Next, gently but firmly bring pressure on the teat with your second finger, forcing the milk down even further. The third finger does the same, then the little finger . . . and if all has gone well the milk will have no place to go but out of the teat.

The first squirt from each udder should be directed into a strip cup, a cup with a black sieve for a cover. Not only is that first stream high in bacteria, but the black sieve will make any abnormality such as stringy or clotty milk show up. This is an indication of mastitis, which demands your attention. Never use the milk from an animal with any sickness.

Then start milking into your pail, using first one hand then the other. With practice, you'll develop rythm. Keep it up until you can get no more, then massage the udder or "bump" it (as kids do when sucking) and you'll be able to get more milk. This massaging is important, not only because the last milk is highest in butterfat but because if you don't get as much milk as possible the goat will stop producing as much as she is capable of.

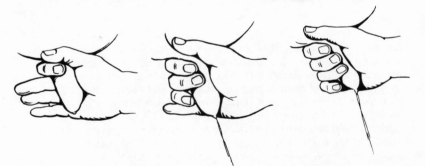

Left. Milk can run out of the teat into the pail or *back into the udder.* So first close your thumb and first finger so the milk can not run back into the udder. *Center.* Next close your second finger—and the milk should squirt out. Discard the first stream—it will be high in bacteria. *Right.* Close the third finger. Use a steady pressure. Don't jerk down.

Upper left. Next close your little finger . . . squeeze with whole hand. *Upper right.* Now release the teat and let it fill up with milk. Repeat the process with the other hand . . . *Center left.* When the milk flow is near to stopping, nudge the bag to see if the doe has let down all her milk. *Center right.* The final bit of milk may be stripped out. Take teat between thumb and first finger. *Lower left.* Now run down length of teat. Milk high in butter fat usually comes at end of milking. But prolonged stripping is bad for the teats and udder. *Lower right.* Strip cup: the first milk is milked into the strip cup. If the milk is "lumpy" it will not pass through the strainer.

There are problem milkers. If you've learned to milk with decent animals you can figure out how to cope with them, but if you're unfortunate enough to have to learn on a trouble-maker, some of the fun will go out of the experience.

First fresheners are most liable to be the culprits, although many older does sometimes develop ornery habits, especially if they know you're using them to practice on. First fresheners are also likely to have small teats, which makes milking difficult, especially for a man with large hands. On some it's possible to milk by using the crotch of the thumb; others will require using the tips of the thumb and index finger in what amounts to stripping.

An occasional doe will have a tendency to kick. This general-ly indicates something is wrong. She's bothered by flies or lice, or you pinched her or your fingernails are too long. Placing the bucket as far forward (away from her hind legs) as possible will help in this situation, and you can lean into her with your fore-arm to help control movement. Leaning into the goat with your shoulder, holding her against the side of the milking bench or wall, will also help to control movement of nervous or ornery animals. This also helps in mild cases of "lying down on the job." Goats that have been nursing kids are especially prone to this sort of unhelpful behavior.

Goats can be milked by machine, but the time spent in clean-ing equipment limits its use to herds considerably larger than any homesteader . . . or sensible beginner . . . would have.

All equipment that comes into contact with milk must be scrupulously clean of course. Milk is not only a highly perish-able product, especially raw milk, but it is also extremely deli-cate. Milk must be cooled *immediately* after milking. It should not be left standing around while you finish up chores. Small containers may be cooled in the refrigerator, but anything more than a quart will not cool rapidly enough for good results unless it is immersed in ice water. Ideally milk should be cooled down to 38 degrees within an hour after milking. That's quite a rapid drop when you consider it was over 100 degrees when it left the udder. Home refrigerators just aren't cold enough.

Cleaning milking utensils is quite different from ordinary household dishwashing. A dishcloth will not clean microscopic pores that hold bacteria that will spoil milk or give it a bad flavor: a brush must be used. Household soaps and detergents contain perfumes that will leave a film on equipment and may cause off flavors in the milk. Household bleach isn't pure enough for the dairy. You can't even use tap water to rinse milking equipment or towels to dry it, because of the bacteria they contain.

There are four dairy cleaning agents. Two are for washing: alkaline detergents and acid detergents. Iodine and chlorine compounds are used for sanitizing.

The alkaline detergent is the basic cleansing agent. However it leaves a cloudy film called milk-stone which harbors bacteria. To get rid of the milkstone you must use an acid detergent, which doesn't have the cleaning power of the alkaline detergent. Most dairy farmers scrub their equipment with alkaline detergent for six days, and on the seventh, when everybody else is resting, they scrub with acid detergent. If you have hard water that makes milkstone develop faster, you could put the acid detergent in the rinse water every day.

Chlorine compounds are used to sanitize equipment, but they're too strong to use for washing udders. There you need iodine compounds. Actually iodine compounds can also be used to sanitize equipment if you measure carefully and let the equipment soak long enough. At least five minutes is required.

Measure any of these materials carefully. If the solutions are too weak they won't do the job they were intended for; if too strong you're wasting money and run the risk of contaminating your milk.

Never let milk dry in a pail or other piece of equipment. Rinse it as soon as possible with cold water, then wash in warm —not hot—water and dairy detergent. Rinse in plenty of warm water with the sanitizing compound, and invert on a rack—not a shelf—to air dry. Do not use a towel.

Commercial dairies *must* follow these procedures. The number of homesteaders who go through all this is open to question. I've described them not because you'll croak from drinking milk that was not produced under hospital conditions, but so those who *want* to do a professional job will know they aren't doing it with soapy dishwater and a dishcloth and towel. If you ever encounter "bad" tasting milk, your milk handling and equipment cleaning procedures are the first things to examine.

14

Records

Record-keeping is necessary for the commercial goat dairy-man, because only through accurate and complete records will he know if he's making a profit . . . or if not, why not.

Record-keeping is a necessity for the fancy or show-ring goatkeeper because only through accurate and complete records will he be able to upgrade his goats to the blue ribbon status he hopes for.

Most homesteaders shun records because they aren't concerned with profit or upgrading, and they think the work is a boring waste of time. And they're wrong on three counts.

It's true the home dairyman doesn't depend on his goats for a living as the commercial goat farmer does. But his profit shows up in milk and dairy products that are cheaper and better than can be purchased in the supermarkets. And even if the homesteader has no intention of ever showing a goat or even coming close to a goat show, he still must know certain facts about his production and management and breeding practices.

Best of all record-keeping can be fun! It becomes a challenge to have does that produce better than their mothers and it's satisfying to look back on records that are several years old and see, in black and white, how you've progressed. No livestock breeder of any kind can afford to be without good records to use as a management tool.

If you have registered goats, pedigrees and registration certificates will be an important part of your files. The person you buy registered goats from will help you with these.

Of more interest to the homesteader, perhaps, are barn records. This is basically a chart showing how much milk each goat produces. A sheet of paper with the goats' names across the top and the days of the month down the left margin works fine. You can write the morning's milk in one corner of each square or imaginary square, and the evening's milk below it, as 4.5/4.

Milk is measured by weight rather than volume, primarily because freshly-drawn, unstrained milk foams, and it's difficult to gauge actual production, but also because it's much simpler and more accurate to deal in pounds and tenths than fractions of quarts. For all practical purposes, a quart of milk weighs two pounds.

It's a good idea to use this sheet to make notations of relevant data. For example, if you note "Susie in heat," you will be alerted to watch for her next cycle in 21 days. Notes on changes in feed, unusual conditions such as a doe not feeling well or any other factor that might contribute to differences in milk production can be a big help in interpreting your records later. It's convenient to note breeding dates and name of the buck, and freshening dates with all pertinent information right on this same sheet. When we used to buy feed instead of growing our own, that went on the sheet too. At the end of the year I had 12 pages with a complete record of the input, output and interesting happenings in my barn.

One of the primary advantages of such a system is that it overcomes the natural forgetfulness of most human brains. Let's face it: few people, if any, are going to remember the statistics from 730 milkings a year which, if the herd consists of three goats with lactations of 305 days, means 1,830 separate entries each year.

In addition to not being able to retain all those statistics, the human brain can distort them. For example, you might be impressed by Susie's production of a gallon of milk in one day and consider her the best in your herd. But your records may show you that a less spectacular producer that just chugged along less dramatically but with a long and steady lactation, actually produced more than the flashing star. If you had to cull milkers or make a decision about whose daughter to keep you might make the wrong choice without records. Such information can also be used to choose the buck to breed each doe to, if you have information on the buck's daughters.

In many cases breeders will note that a relatively few top does produce as much as a much larger number of poorer producers. Since poor producers require just as much work as good ones and eat virtually as much, it follows that milk from the lower third or half of your herd costs more than milk from the top half or two-thirds. This suggests that you could get more milk for the same amount of time, effort and money, by replacing the poor doers with daughters from the best does. If you don't need that much milk you might be able to eliminate one or more animals, increase your herd average, and thus reduce your feed bill.

Records of breeding, expenses, income and milk production are all basic, and really don't take much time or knowledge of accounting to keep. Just the act of keeping them will tell you a few things about your goatkeeping operation. But it's also possible to squeeze a lot more helpful and interesting information out of those records.

For instance, you might want to know how much your goat milk costs you, either to set a fair price on milk you might sell or to decide if you can afford goats or should go back to the grocery store dairy case to save money. (In the latter case your records will also show you *why* your milk is too expensive, which is what you must know to correct the situation.)

Here are some actual records from a herd of six does. You can tell they're old from the milk prices—but then feed costs were lower, too.

Income

Milk sales at 50c a quart	$120.75
Sales of stock	260.00
Family milk at $1 a gallon	248.00
Stud fees	130.00
Boarding fees	159.50
Misc. (Disbudding services)	12.00
	$930.25

Expenses

Purchase of stock	$ 55.00
Feed—grain, minerals, salt	431.20
Hay	240.25
Veterinarian and medicine	67.85
Repairs on equipment	78.17
Fencing	36.00
Supplies	89.29
Advertising	59.80
Registry, transfers, etc.	50.00
Telephone	17.19
	$1124.75
Expense over income	$ 194.50

This balance sheet tells us the goats are a losing proposition here, but to learn why we must go to other records.

In the first place, total milk production for the six does in this herd was 5,900 pounds. That's a poor average. Two does came in with little milk and were dried off after a few weeks, but were kept on the payroll as boarders. There may have been good reasons for their lack of production, but statistically speaking they should have been culled. The bottom line of the balance sheet would have come out looking better.

The records show this was a poor kidding year. There were eight bucks and four does, and one of the does died. Over the years any given herd will produce bucks and does pretty much on a 50/50 basis, and the sale of additional doe kids, especially if purebred, would also have helped the balance sheet.

Medical expenses seem high, although they go much higher in some herds. Others have none. Checking into the nature of the problems might indicate a condition that could be corrected by different feeding or management practices, or if the vet bills apply to one or two does or to a particular family, culling is in order.

Fencing and the purchase of stock should rightfully be treated as capital expenses. These are apportioned over their useful lifetimes. Other examples of this would be milking equipment, pens, or a milking stand.

It can be seen that having these records and using them is the only way to make intelligent management decisions . . . the only way to lower your milk bill.

With good records you can tell what each goat is worth. Just add up her annual production, put a price tag on it, and add in any income from kids. (To the homesteader, "income" may be in the form of meat, or the worth of a replacement doe.)

Figure your capital costs, money you spent on things that aren't used up all at once. This includes milking equipment, feed pans and water buckets, fencing, tools and other equipment, and the goat itself. Naturally you don't charge off all of this in the first year. The milk pail might last 20 years: take one-twentieth of the price. The goat might be good for another five years. Take one-fifth of what you paid for her. Go down the line of capital equipment in that manner, and add up capital costs for one year.

Then add up your operating expenses: hay and grain, electricity in the barn, veterinary fees, milk filters and anything else that was purchased and used up.

By adding up the operating expenses and one-year cost of capital equipment and stock, subtracting that from the value of the goat's production, you'll know her annual value to you. By adding up all costs and dividing by the number of quarts of milk produced you'll know the actual cost of your milk.

This isn't really accurate, of course, but it's adequate for homesteading and much better than the complete neglect of such accounting on homesteads. And if you're inclined, you can figure in labor, the value of manure and return on investment.

If more goat raisers kept such records, you can be very sure there wouldn't be so many $10 goats for sale! More people would pay more attention to culling and proper management, too, if they knew what their goats were costing them.

One other point is important here. An examination of actual records frequently shows that the herd breaks even not because of the value of milk, but because of the value of the kids. It becomes readily apparent that a purebred and registered herd that can command top price for its animals will come out far ahead of a herd of grades whose kids are a drug on the market. Purebreds *do* have something to offer homesteaders even if they aren't interested in showing.

15

Chevon

The meat of goats is called chevon, from the French word for goat, *chèvre*. That's not being pretentious. Not too many people would be interested in eating dead pig, but they eat pork, from the French word for pig which is *porc*.

Chevon is very popular in certain cultures, in this country especially among people of Spanish-American, Greek, and Jewish heritage. Kid is an important part of the meals for spring festivals of Easter and Passover. More than 200,000 goats are slaughtered annually under federal inspection, mostly in the Southwest where they can be raised cheaply on range and where a market exists.

In most of the country dairy goats raised in confinement can't be considered meat animals because of the labor and expense involved. They are, however, an important by-product of dairying. Over the years any farm will average 50 percent buck kids. They obviously can't all be kept. While there is a limited demand for wethers (castrated males) as pets in some areas, it is probably more merciful in most cases to butcher them for meat.

In addition to unwanted males, any dairyman—cow or goat —will have cull or aged does that simply are not paying their way. Resist any temptation to sell them as milkers to someone else. This is a hard, hard thing, and I'm positive I'm not the only goat keeper who has kept a favorite old doe around til she died of old age because I didn't have the heart to dispose of her. There's nothing wrong with that just so long as we understand such sentiment is a luxury, and often an expensive one.

As this is being written the public outcry in reaction to the slaughter of young calves is still fresh in my memory. If you are one who had difficulty appreciating the farmers' position on this, the entire subject of excess males is going to be an interesting one. Keep them all . . . at a cost of $100 or more a year each . . . and your goats will soon be a drain on the budget rather than a help. Sell them as pets and you practically guarantee them a life of abuse, neglect and misery, if not from the first owners then from subsequent ones, because male goats seldom remain as pets at one home for long. If you can't face the thought of killing or eating goats, and can see the problems just mentioned, you can have a more hardened neighbor destroy them and bury them . . . which is what the farmers did with the calves. Or you butcher them and save money on your meat bill.

The choice is yours. But, you *will* have to make it. And the time to give it some consideration is before you even begin raising goats.

Goats are commonly slaughtered at four different stages. Most popular for the Easter-Passover market are milk fed kids weighing between 20 and 30 pounds. On homesteads where milk is valuable (and also to avoid butchering animals that you've become somewhat attached to) it's possible to butcher kids at birth and dress them out like rabbits.

Many homesteaders who make a conscious effort to produce meat from their dairy herd castrate buck kids and feed them out for 6-8 months.

And finally, cull does or old animals can be processed into jerky, salami, or anything that makes use of meat that isn't especially tender. (Bucks that are being culled should be castrated about two months before butchering and fed a liberal grain ration.)

If you dispatch the animal with a gun, aim from the back so as not to frighten it by the sight of the barrel. Many people prefer to use a hammer. A sharp blow to the skull will render the animal unconscious, then cut the jugular, with a sharp, stout knife.

Thorough bleeding is important for the pale meat preferred in our culture. Do not damage the heart in any way so it may continue to pump blood, and hang the animal head down to allow complete drainage. If you do much butchering a gambrel hook will prove useful, although a carcass can be hung by passing a metal or wooden rod through the tendons or even by tying it to a tree branch or rafter with a rope.

The Greeks, who have a good deal of experience in goat butchering, cut a small incision between the hind legs and blow up the hide like a balloon. This helps separate the hide from the meat and makes skinning easier and cleaner. Most people would rather use a tire pump than their mouths, but perhaps the best method of all is to insert the nozzle of a garden hose into the incision and fill it with cold water. The effect is the same, and in addition the cold water helps cool the meat.

To skin the animal, carefully cut a slit from between the hind legs to the throat. Don't cut too deeply: once started you can usually work your fingers beneath the skin to hold it away from the body and the cutting will go faster.

From the ends of this cut continue out along the insides of all four legs. The skin is tighter on the legs and again, be careful not to cut into the meat!

The pelt will be attached at the tail. To remove that, cut around the anus and loosen it until a length of colon can be pulled out. Tie off the colon with a piece of strong string to avoid possible contamination. Cut it off above the string and let it fall back into the body cavity. Then cut off the skin at the base of the tail.

If you're interested in eliminating waste and making your homestead as efficient as possible, save the tail for stew meat.

If you're saving the hide cut it off as close to the ears as possible. Skins from newborn goats are more like fur than hide, and many useful items can be made from any tanned goat skin. If you aren't interested in the hide you can cut the head off with it. In either case remove the head at the base of the skull.

Next cut down the belly from the hole already made, to the brisket. If the animal was starved for 24 hours before butchering the paunch will be empty and there is less chance of cutting into it, but use care anyway. Let the paunch and intestines roll out and hang. Work the loosened colon end down past the kidneys and carefully remove the bladder. Pull out the liver and remove the gall bladder by cutting off a piece of the liver with it. If the gall bladder breaks and spills bile on the liver wash the meat in cold water immediately to avoid a bitter taste that will result.

The offal will fall free when the gullet is cut. Saw the brisket (an ordinary wood saw will work if you don't have a meat saw) and remove the heart and lungs. Clean out any remaining pieces of tissue, wash the carcass with cold water, and wipe it dry.

The skull can be split to get the brains, and the tongue removed. Wash the liver, heart, and tongue in cold water and hang them to dry.

Newborn goats weighing about seven pounds can be cut up like rabbits. Cut through the back just in front of the hind legs and just behind the front legs. The saddle can also be cut into two pieces and the legs should be cut apart, giving you six pieces of meat.

Larger animals, cut like lamb, will yield roasts, chops, and trimmings that can be used in patties or mixed with pork for sausage. Some of the larger pieces, such as the legs, can be cured like hams or corned. A good source of information on cutting and curing lamb, (which applies to chevon) is "A complete Guide to Home Meat Curing" available from Morton Salt Company. Parts of the booklet were also reprinted in both Countryside & Small Stock Journal and Mother Earth News and reprints of the lamb section are still available from the former as of this writing. Lacking this information . . . or even with it, if you attack the project the way some of us do, you simply cut off pieces that "look about right to be a roast" or whatever. After a few times you'll develop a definite method. There really isn't anything you can do wrong at this stage except not end up with the "proper" cuts as sold in the supermarket. Yours will be every bit as edible, and probably even more delicious.

If you like lamb you can use your favorite lamb recipes with chevon. There are also many ethnic dishes from such areas as Greece and Turkey that call for chevon. Oregano is a good spice to use with chevon and the meat is excellent in curries.

16

Goat Milk Products

One of the joys of having goats is the dairy products you can make in your own kitchen. At certain times of the year, especially, you will have a surplus of milk that can be turned into a variety of products that will make you more independent of the supermarket and will make your goats more valuable. There's no sense getting milk that costs you 10 or 20 cents a quart and then throwing out half of it because you can't drink it all!

Milk can be frozen, of course, and if you know you're going to run short later maybe this is the first course to consider. Freeze it in plastic jugs, and leave an air space for expansion. Thawed frozen milk is somewhat watery, and while it's fine for cooking we prefer to mix it with fresh milk for drinking.

Cheesemaking

The next step is cheesemaking. This doesn't require much equipment, it takes very little actual working time and it can get to be quite a hobby. There is as much art to making cheese, however, as there is to making wine. Don't expect to come up with any "vintage" cheeses without experience . . . or luck. (Neither helps me much, although luck seems more important than experience.)

141

For beginners, all you really need that you don't already have in your kitchen is rennet, a cheese thermometer and cheesecloth. Rennet is available from several mail order firms, as are floating dairy thermometers.

Cottage Cheese

The easiest cheese to make is cottage cheese. Start by warming one gallon of milk to 86 degrees. Add rennet (follow the directions that come with it) and let the milk stand in a warm place until a curd forms, usually an hour or so. When that happens (and the timing isn't critical), cut the curd with a long thin bladed knife. By cutting it in one-half inch squares on the surface, and then angling the knife to make similar slices at right angles to the first ones, you'll end up with small cubes. Stir the mass . . . but very gently . . . to make sure you got all of it, then slowly warm the curds and whey to 110 degrees. The heating should be very slow and gentle. Within limits the longer it heats the firmer the curd will be.

When it suits your taste, pour the curds and whey into a colander lined with cheesecloth and let it drain for a few minutes. Then run cold water over it to rinse off the whey.

You can eat it like that or add salt if you wish or add cream. Even people who don't like cottage cheese (from the store) will appreciate this. I hear it keeps for a week in the refrigerator, but set on the table ours doesn't last nearly that long.

It is possible to make cottage cheeses (and others) without rennet. Some recipes call for adding cultures or a starter, buttermilk, sour milk or even yogurt. (See "Making Homemade Cheeses and Butter" by Phyllis Hobson, available from Garden Way Publishing.)

Actually, in spite of all the recipes for cheese available, the fact that my family loves cheese, and the fact that I love to experiment in the kitchen and have made hundreds of cheeses, I stick with one basic recipe. One reason is that I like it, but perhaps more importantly, even though I've been making it for years it seldom turns out the same twice so I don't really *need* more than one recipe. (I don't profess to be a cheesemaker, obviously.)

The cheese is affected by the length of time you heat the curd, how hot it gets, the amount and duration of pressing and humidity. I find it very difficult to make hard cheese during the summer in Wisconsin, for example.

This recipe is the basic one used for almost all cheeses, so if you master this better than I have, and want to experiment further, get Phyllis Hobson's book and you'll already know the basics.

Warm the milk to 86 degrees in a large kettle. Since it takes ten pounds of milk to make one pound of cheese, I don't like to mess around with less than six quarts of milk, but that's up to you.

Add the rennet. One-quarter tablet will suffice for small amounts of milk. It is dissolved in cold water and thoroughly stirred into the milk with the heat turned off. Let it stand until the curd forms.

When the curd is firm enough to be lifted up by a finger it can be cut. Using a long knife (we have a bread knife which seems to be ideal) cut the curd into squares. Cut at one-half inch intervals vertically, then at right angles to that on the diagonal. Stir it carefully to find any pieces you missed, and cut them so all pieces are uniform.

Then reheat the cheese-to-be . . . *very* slowly. The temperature shouldn't rise more than two degrees every five minutes. Heat it to 100 degrees, and hold the temperature there until the curd is as firm as you want it. You'll have to stir it once in a while to keep the curds from lumping together, but do it gently. It takes $1^{1}/_{2}$ to $2^{1}/_{2}$ hours for the curd to get to the right consistency.

If the curd is not firm enough when you remove it the cheese will be pasty and may sour. If the curd is too firm it will be dry and crumbly.

When it's ready, pour the curd and whey into a large container lined with cheesecloth. Lift the cheesecloth out and let the curds drain by hanging it over a pan. (I've devised a simple method of twisting the ends of the cheesecloth around the handle of the original cheese pot and tilting it so the whey collects on the bottom side near the opposite handle. This saves hanging the bag from a cupboard door handle and dirtying an extra pot to wash.)

When it stops dripping, place the curd in a container, break it apart and add salt.

The whey makes good feed for chickens or pigs, but don't give it to kids or goats. It will cause scours.

Notice that this is the same method used for cottage cheese. The pressing is what makes other cheese different.

You can make a simple cheese press from wood or from a large coffee can and a few pieces of wood, but for a start neither of these is necessary. Merely shape the cheese into a ball and make a "bandage" or headband from a clean dish towel. You'll need it maybe two inches wide for a small cheese. Wrap this around the cheese and pin it in place with a safety pin.

Place it on a board, and put another board on top of it. It will continue to drain: put it in a large kettle or the sink. Weight down the top board with a brick or anything else that's heavy. For hard cheese, pressing pressure must be much greater, and some home cheese makers have lever arrangements that allows

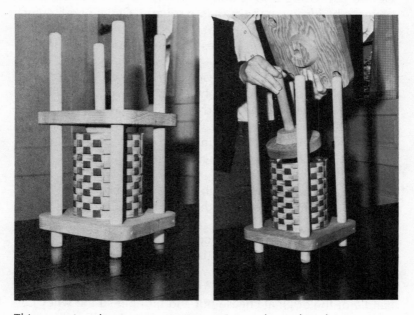

This very nice cheese press was made from a large dowel, a few pieces of scrap lumber, and a coffee can covered with straw matting.

pressure to be applied without stacking up 30 or 40 pounds of bricks. (Bricks stacked up tend to fall over anyway when the cheese starts to settle.)

The cheese must be turned several times a day. Generally speaking, the longer it's pressed and the greater the weight, the harder the cheese. There are specific times and weights for various recipes, so you can't really go wrong. If you use anything handy for a weight and turn it and press it according to your own lifestyle you might end up with a cheese that's different from someone else's, but it'll still be cheese.

When it's pressed, wipe the cheese with a clean dry cloth and check it for cracks and pores. Seal these by dipping the cheese in warm water and smoothing them over with your finger.

Put the cheese on a shelf in a cool, dry place. The ideal is 56 degrees, too warm for the refrigerator, and too cool for most other home locations except at certain brief periods of the year in most areas. Again, this is no problem, but it will affect the finished product. Turn it daily or the bottom will mold.

In a few days a rind will begin to form. It should then be paraffined to prevent it from drying out completely. A half pound of wax is enough for a small cheese. Melt the wax, hold half the cheese in it, remove it and let it cool. Do the other half and check your work for leaks.

The cheese goes back on the shelf, but must still be turned daily. It can be eaten as is, of course, but the length of curing also affects the finished product. A mild cheese will be ready in about six weeks; a cheese cured for six months will have a sharper flavor. Curing takes longer at lower temperatures, but spoilage occurs if the temperature is too high.

As you can see, there are many variations possible with this basic recipe, and by controlling your cheesemaking processes (and keeping records of what you did) you will probably find one that suits you. And if you really get into cheesemaking there are many other possibilities.

For example, cheddar is made by following the basic directions up to pouring off the whey. Then the cubes of curd are placed in a colander and heated to 100 degrees (in the oven or a double boiler) for 1½ hours. When the curd forms a solid mass, instead of the cubes you started with, slice it into one inch thick strips. Turn these every 15 minutes to allow them to dry evenly, still holding the temperature at 100 degrees. After an hour go back to the basic directions beginning with salting the curd. Cheddar takes six months to cure.

We like a variation of what must be a form of feta. When the basic cheese is pressed, instead of curing it, cut it into cubes about two inches square and place these into jars of brine. Eat them right out of the brine.

The possibilities are endless. But alas there are certain cheeses that can't be duplicated at home. Limburger, camembert and certain others require special cultures which are closely-guarded secrets of the cheesemakers. But there are enough others to keep any goat owner happy.

Yogurt is another product popular among goatkeepers. Homemade yogurt is so superior to the supermarket variety that there's no comparison. There are several ways of making it, with everything from a special yogurt maker to a heating pad to using solar energy.

Warm the milk to 100-110 degrees. Add culture (you can buy cultures, or use one cup of store-bought yogurt . . . and save a cup of your homemade for the next batch. It will "wear out" after a while though, and you'll need to use new culture again.)

A yogurt maker will keep the milk at 100 degrees automatically. But you can also put the warm mixture into a pre-heated thermos and wrap it in towels to help hold the heat in. Or you can set a casserole dish in a warm oven and leave it overnight with the heat off. Or you can use a heating pad, the back of the old wood stove, or set a glass-covered container in the sun on the right kind of day. It will take 5-6 hours at 100 degrees.

Butter is difficult to make from goat milk, but only because of its "natural homogenization". If you have a cream separator there's no problem, but without one you'll have a hard time getting enough cream to warrant winding up the churn.

One method of getting cream from goat milk, or some of the cream, is to leave the milk in flat pan with as much surface exposed to the air as possible. You'll be able to skim off some of the cream in about 24 hours. This cream will keep for a week in the refrigerator, and in a week you'll get enough to make butter.

The cream should "ripen", either by sitting in the refrigerator a week while you accumulate enough, or by leaving it at room temperature for a day. Then put it into the churn, but don't fill it more than half full. Or you can use an electric mixer, a French whip or even just shake it in a covered fruit jar. It should take about 20 minutes.

When the butter comes, drain off the buttermilk (this isn't cultured buttermilk but it's drinkable and good for cooking).

This simple butter churn was made from an egg beater.

An idyllic goat scene.

Now you have to work the butter. Use a spatula or wooden spoon, pressing the butter against the side of the bowl and pouring off the buttermilk that is pressed out.

Then wash the butter in cold water to get out any remaining traces of milk, which will cause the butter to spoil. Repeat the washing until the water comes off clear.

Most people prefer their butter salted. Add salt according to taste and work it in.

When you get really good at making cheese, yogurt and butter, and use fresh delicious goat milk in custards, baking and other cooking, you may find you don't have enough milk left to drink! The solution? Simple! Go back to the beginning of the book . . . and get more goats!

For Further Information

There are a great many local and regional dairy goat clubs and associations, many of which have excellent newsletters as well as educational programs. But since they are run by elected officers and have no permanent addresses, it isn't feasible to list them here. See if your county agent knows of any such clubs near you.

You can also check with the agent on the existence of any 4-H dairy goat projects in your area. These are becoming increasingly popular.

The two dairy goat registry associations which sponsor shows and offer other benefits to dairy goat raisers are:

American Dairy Goat Association, Spindale, NC 28160

American Goat Society, 1606 Colorado St., Manhattan, KS 66502

In Canada, registration of goats is handled by the Canadian Goat Society, Canadian National Live Stock Records, Holly Lane, Ottawa, Ontario K1V 7P2.

Two national magazines regularly print information on goats. *Countryside & Small Stock Journal,* Waterloo, WI 53594, has a regular goat section along with information on other small-farm projects. *The Dairy Goat Journal,* Box 1908, Scottsdale, AZ 85252, covers goats exclusively, and has more to offer those interested in showing registered goats.

Books on goats include the following:

Goat Husbandry, by David Mackenzie. The most complete goat book available. About $18. Try bookstores in your area.

Dairy Goats—Breeding, Feeding, Management: Good on reproduction, kidding, and building plans. This booklet is available from ADGA and *Countryside.*

The dairy goat raiser is usually involved in many other homesteading activities and finds that an up-to-date personal library is essential to provide the information he needs at his fingertips.

There are many good books available; here are some excellent choices.

Raising Rabbits the Modern Way, by Robert Bennett. Everything for the home and commercial producer. 160 pp. illus. $7.95 + $2.00 P & H. Order #067-5.

The Family Cow, by Dirk van Loon. Covers cow buying, handling, housing, feeding, milking, caring, breeding, and calfing. Plus information on land use, hay and tools. 272 pp. illus. $8.95 + $2.00 P & H. Order #066-7.

Raising Poultry The Modern Way, by Leonard Mercia. In addition to stock selection, brooding, rearing and more, you are given current methods of disease prevention and treatment for laying flock, meat chickens, turkeys, ducks and geese. 240 pp. illus. $8.95 + $2.00 P & H. Order #058-6.

Keeping the Harvest: Home Storage of Fruits and Vegetables, by Nancy Thurber and Gretchen Mead. Buy this one and forget the other food storage books. 216 pp. photos & illus. $9.95 + $2.00 P & H. Order #247-3.

Raising Your Own Turkeys, by Leonard Mercia. Complete, up-to-date, how-to information on raising turkeys, from young poults to delicious, thick-breasted birds. 160 pp. illus. $6.95 + $2.00 P & H. Order #253-8.

Down-to-Earth Vegetable Gardening Know-How, by Dick Raymond. Learn how to double even triple your yield. A treasury of vegetable gardening information. 144 pp. charts, photos, & illus. $7.95 + $2.00 P & H. Order #271-6.

These books are available at your bookstore, lawn & garden center, or may be ordered directly from Garden Way Publishing, Dept. 4412, Schoolhouse Road, Pownal, VT 05261. Send for our free mail order catalog.

Index

Abortion, 82
Abscess, 83
Advanced registry, 27
Age, 30
Afterbirth, 112
Alfalfa, 56, 63
American, 21
Artificial insemination, 103

Bang's disease, 85
Bedding, 34
Billy goat, 1
Bloat, 85
Bottle feeding, 118
Breed improvement, 94
Breeding, 101
  season, 1
Breeds, 3
Browsing habits, 2
Bucks, 93
Bulk (in diet), 57, 58
Butchering, 138
Butter, 146
Buying goats, 19

Calcium, 51, 66
Castrating, 79, 120
Cheese, 141
Chevon, 137
Clipping, 79
Colds, 86
Colostrum, 116, 117
Comfrey, 65
Concentrates, 57
Conformation, 22
Cuts, 86
Cystic Ovary, 86

Dehorning, 74, 77
Detergents, 129
DHIA, 27

Digestive system, 46
Disbudding, 71
Drying off, 101

Enterotoxemia, 87

Feed, 45
  changes, 49
  composition, 53
  formulas, 59
  home mixed, 46
  organic, 46
  protein content of, 51, 54-55, 57
  quantity, 57, 67
  roughage, 53
  trace minerals, 65, 51
  vitamins, 52
  weights, 58
Fencing, 37, 43
French Alpines, 4

Gallon milker, 10
Goat pox, 86
Goat prices, 29
Grade goats, 20
Grooming, 69

Hay, 48, 52, 53, 56, 63
Health, 81
Heat periods, 102, 104
Hermaphrodites, 104
Home grown feed, 63
Hoof trimming, 69
Horns, 26, 74, 77
Housing, 33

Johnne's disease, 87

Ketosis, 58, 91
Kidding, 109

EAU CLAIRE DISTRICT LIBRARY

Kid raising, 117
Kids, feeding, 119

Lactation curve, 9
Lactation period, 10
La Mancha, 7
Lice, 88

Mange, 90
Mastitis, 88
Maturity, 105
Milk, 9
  care of, 128
  digestibility, 15
  fat globule size, 15, 16
  off flavor, 16, 129
  specific gravity, 15
  testing, 16
  weight of, 27
Milkers, problem, 128
Milk fever, 88
Milkhouse, 41
Milking, 121-128
  equipment, 121
  procedure, 124
  sanitation, 121
  time, 121
Milk replacer, 119
Minerals, 51, 65
Molasses, 58

Nubians, 3
Nursing, 118

Pan feeding, 58
Pedigrees, 20
Phosphorus, 51
Poisoning, 65, 88
Poisonous plants, 65, 88
Production records, 4, 10, 11, 12,
    26
Protein, 49, 51, 57
Purebred, 8, 20, 135

Pygmy, 7

Ration, 59
Recorded grades, 21
Records, 131
Registration, 20, 22
Roughage, 48, 52, 53, 56, 63
Rumination, 46

Saanen, 5
Salt, 57
Scent glands, 1, 98
Scours, 81, 48
Soybeans, 62
Space requirements, 2, 36
Spotlight sale, 29
Staking out, 3
Star milker, 28
Sterility, 104
Supernumerary teats, 113

Tattooing, 77
TDN, 52
Tetanus, 90
Toggenburg, 5

Urea, 62

Vitamins, 52

Water, 39, 60
Wattles, 25
Weaning, 120
Weeds, 63
  poisonous, 88
Weight,
  of goats, 2
  of milk, 27
  of feeds, 58
Worms, 91

Yogurt, 146